Mark's copy

COMITE EURO-INTERNATIONAL DU BETON

DURABLE CONCRETE STRUCTURES

DESIGN GUIDE

Thomas Telford

Major contributions to this Design Guide were made by the following members of CEB General Task Group 20: Durability and Service Life of Concrete Structures

S. Rostam (Reporter), Copenhagen, Denmark
R. F. M. Bakker (from December 1984), Ijmuiden, The Netherlands
A. W. Beeby, London, Great Britain
G. Hartl, Vienna, Austria
D. Van Nieuwenburg, Gent, Belgium
P. Schiessl, Aachen, Germany
L. Sentler, Lund, Sweden
A. P. van Vugt (until December 1984), 's-Hertogenbosch, The Netherlands

Published by Thomas Telford Services Ltd, Thomas Telford House, 1 Heron Quay, London E14 4JD, UK, for the Comité Euro-International du Béton, Case Postale 88, CH-1015 Lausanne, Switzerland

Second edition 1989
Reissued 1992

British Library Cataloguing in Publication Data
Durable concrete structures
I. Comité Euro-International du Béton
624.1

ISBN 0 7277 1620 4

Typeset by MHL Typesetting Limited, Coventry
Printed and bound in Great Britain by Butler & Tanner, Somerset

Preface

Concern in recent years over the occasionally inadequate durability of concrete structures has led to intensified research into the causes and nature of degradation processes, and to the development of general strategies for handling such situations. Since the late 1970s the CEB has been active in solving the technical aspects of premature degradation of concrete structures. This Design Guide has been prepared by the General Task Group No. 20: Durability and Service Life of Concrete Structures.

The Guide is a synthesis of four previous works by the Group: a State of the Art Report, presented in 1982 as CEB Bulletin No. 148;[1] the international workshop on the subject organized in Copenhagen in 1983 in co-operation with RILEM;[2] the *Draft CEB Guide to Durable Concrete Structures*;[3] and the second international workshop organized in co-operation with RILEM in Bologna in 1986.[4] Valuable comments have been received on the Draft CEB Guide[3] from technical organizations, national delegations and individuals. The Task Group has considered in detail all comments and proposals received, and the results have been incorporated in this Design Guide.

The Guide is intended for practising engineers rather than materials specialists. It presents simplified models of degradation mechanisms and influencing factors. However, these models are believed to be consistent with present-day knowledge of the complicated physico-chemical mechanisms determining the intensity of degrading actions and resulting deterioration mechanisms in concrete structures.

The members of the Task Group are all cordially thanked for their many valuable contributions during the preparation of this Guide, and for their continuous enthusiasm, which has been of mutual inspiration for the work.

Steen Rostam, MSc, PhD
Reporter of the Task Group
Copenhagen, June 1989

Foreword

It became apparent to the Comité Euro-International du Béton (CEB) some years ago that there was a need for definitive information and guidance on the performance of concrete structures during their full service life and on the means to assure the desired level of performance during the design and construction process. Thus the topic of durability and service life of concrete structures was assigned to a general task group for study.

This group, having worked for a number of years, and having disseminated and discussed its work widely, has now produced a design guide aimed specifically at practising designers of concrete structures. Hopefully, it will assist them in creating structures fully fit for their purpose during a defined service life with the minimum of maintenance.

The subject is of considerable interest worldwide and so the CEB aims to ensure the optimum dissemination of its work. It is my pleasure and privilege, as President of the CEB, to commend this guide to the design profession as a beginning to the process of making real design for durability and performance an integral part of the traditional design and construction procedures.

R. E. Rowe
President, CEB

Acknowledgements

This complete CEB Design Guide is the responsibility of the Task Group as a whole. Certain members of the Group have been largely responsible for individual sections of the Guide

A. W. Beeby: 3.1, 7, 8.9, 13.1, 13.2, 13.4
G. Hartl: 3.1, 3.3, 12.1.1.1, 12.1.1.3
D. Van Nieuwenburg: 10.3, 11
S. Rostam: 7, 9, 10, 13.3, 14
P. Schiessl: 2, 3.2, 6, 12.1.1.2, 12.2
A. P. van Vugt/R. Bakker: 4, 5, 12.1.2, 12.1.3

From outside the group, E. J. Pedersen from the Concrete and Structural Research Institute, Denmark, has provided an appendix on curing of concrete structures, at the request of the Task Group.

Discharge of the special obligations laid on the Reporter has been made possible by the valuable support of the Danish Academy of Technical Sciences, the Department of Structural Engineering, Technical University of Denmark, and COWIconsult, Consulting Engineers and Planners AS.

All services and financial support received are most gratefully acknowledged.

Steen Rostam, MSc, PhD
Reporter of the Task Group
Copenhagen, June 1989

Contents

1. Introduction

This design guide attempts to synthesize basic technical knowledge and current engineering experience regarding durability characteristics of concrete and concrete structures, and present these in a practical guide for the design and construction engineer.

Due to the complex nature of environmental effects on structures and the corresponding response, it is believed that true improved performance cannot be achieved by improving the materials characteristics alone, but must also involve the elements of architectural and structural design, processes of execution, and inspection and maintenance procedures, including preventive maintenance.

It should at least be possible for all persons involved in the creation and use of concrete structures to obtain a minimal understanding of the most important deterioration processes and their governing parameters. In certain cases, such basic knowledge is a precondition for the ability to take the correct decisions at the right time when seeking the required durability. Schematic approaches to service life design are not considered reliable in practice.

For these reasons, in the first part of the guide, the theoretical background regarding possible deterioration processes and their governing factors is presented in terms of simplified engineering models. These models are as compatible as possible with the more complicated treatments of the same mechanisms that may be presented on the materials science level. The more directly applicable recommendations are presented in the second part.

A comprehensive treatment of heat and moisture curing is given in Appendix A; the essence of this is outlined in part II.

The fundamental approach adopted in the guide is illustrated in Figs 1.1 and 1.2, which show the interrelations between the main factors influencing durability. It can be seen that the combined transportation of heat, moisture and chemicals, both within the concrete mass and in exchange with the

Fig. 1.1. Relationship between the concepts of concrete durability and performance

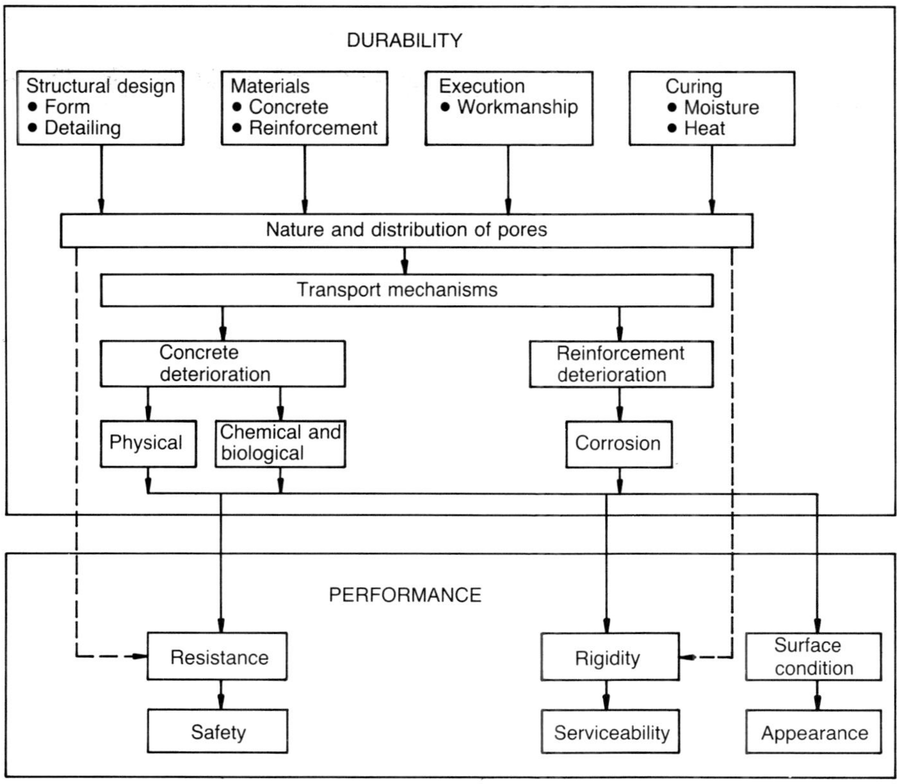

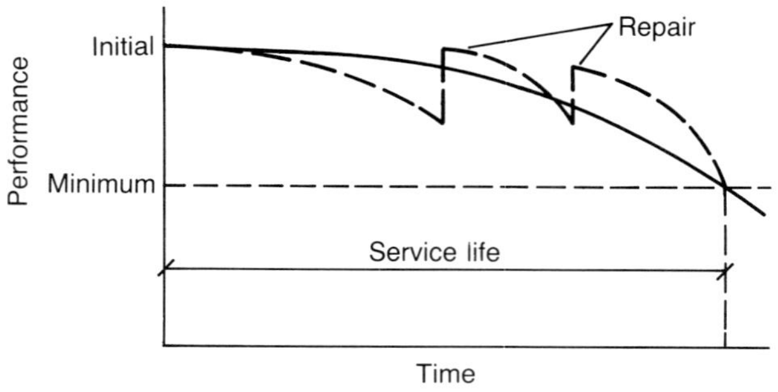

Fig. 1.2. Relationship between concrete performance and service life

surroundings (the microclimate), and the parameters controlling these transport mechanisms, constitute the principal elements of durability.

The presence of water or moisture is the single most important factor controlling the various deterioration processes, apart from mechanical deterioration. The transport of water within the concrete is determined by the pore type, size and distribution and by cracks (microcracks and macrocracks). Thus, controlling the nature and distribution of pores and cracks is essential. In turn, the type and rate of degradation processes for concrete (physical, chemical and biological) and for reinforcing or prestressing reinforcement (corrosion) determine the resistance and the rigidity of the materials, the sections and the elements making up a structure. The surface conditions of the structure are also determined in this way, and this is reflected in the safety, the serviceability and the appearance of a structure; i.e. these processes determine the performance of the structure.

What is of concern in practice is to ensure a satisfactory performance over a sufficiently long period of time. This performance over time — whether due to initial good quality, or to repeated repair of a not-so-good structure — may be termed the service life of the structure (Fig. 1.2).

The modelling of these aspects of durability and service life is covered in an introductory section on transport mechanisms in part I. Part II starts with a section on the classification of environmental conditions affecting concrete — an important element of the problem area, concerning which available information is unfortunately scarce.

A detailed bibliography is given in ref. 1.

2. Transport mechanisms in concrete

2.1. Transport mechanisms: basic considerations

In nearly all chemical and physical processes influencing the durability of concrete structures, two dominant factors are involved: transport within the pores and cracks (Fig. 2.1), and water. Both the transport of gases and the transport of water and dissolved deleterious agents and the binding mechanisms are important. The rate, extent and effect of the transport are largely dependent on the pore structure and cracks and on the microclimate at the concrete surface. In this context, pore structure signifies the amount of pores and the pore size distribution.

The pore structure and crack configuration, and the filling of pores and cracks with water, are determining factors in respect of the transport of water and gaseous and dissolved substances. In addition, the rate of transport depends considerably on the transport mechanism. In the event of chemical binding mechanisms being involved, the chemical composition of the cement and the properties of the aggregates are also of importance. All transport mechanisms are mainly a function of the pore structure and crack configuration, and are determined by the same processes.

The major deterioration mechanisms, the fundamentals of the pore structure of the concrete, the binding mechanisms for water, and the transport phenomena are briefly illustrated by means of three characteristic environmental conditions.

2.2. Pore structure of concrete

In addition to the microclimate, permeation is decisively influenced by the pore structure of the cement paste. For a characterization of the open pore structure with regard to the transport of substance into and within porous

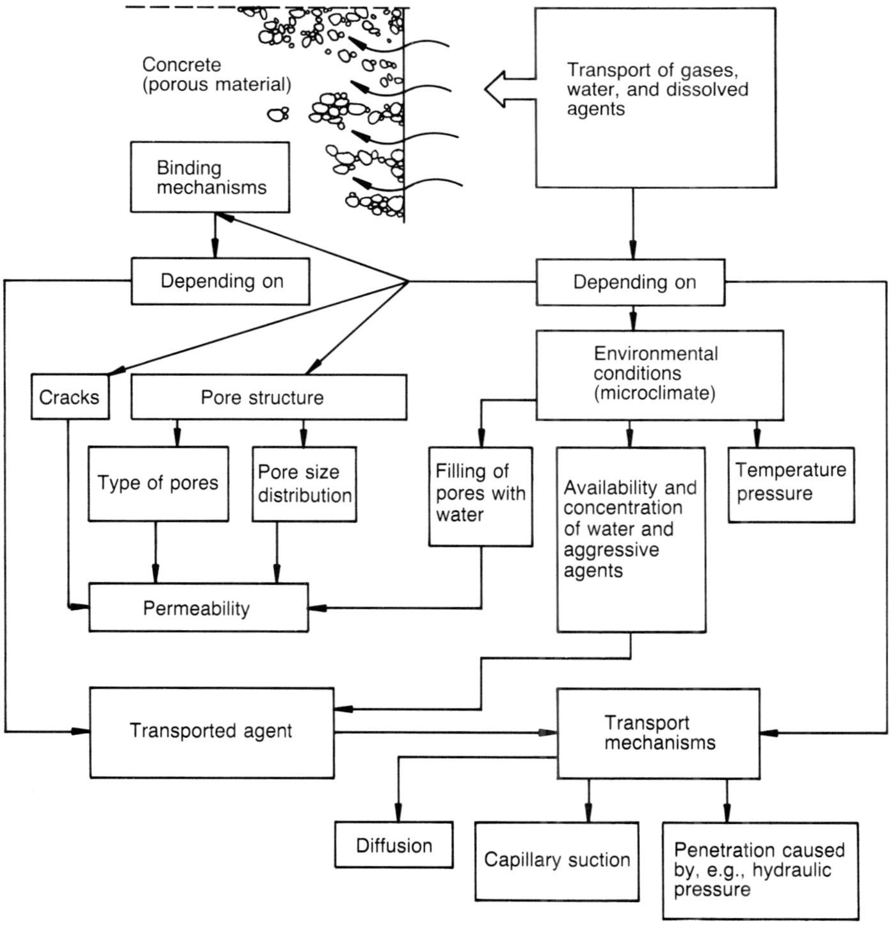

Fig. 2.1. Transport phenomena in concrete

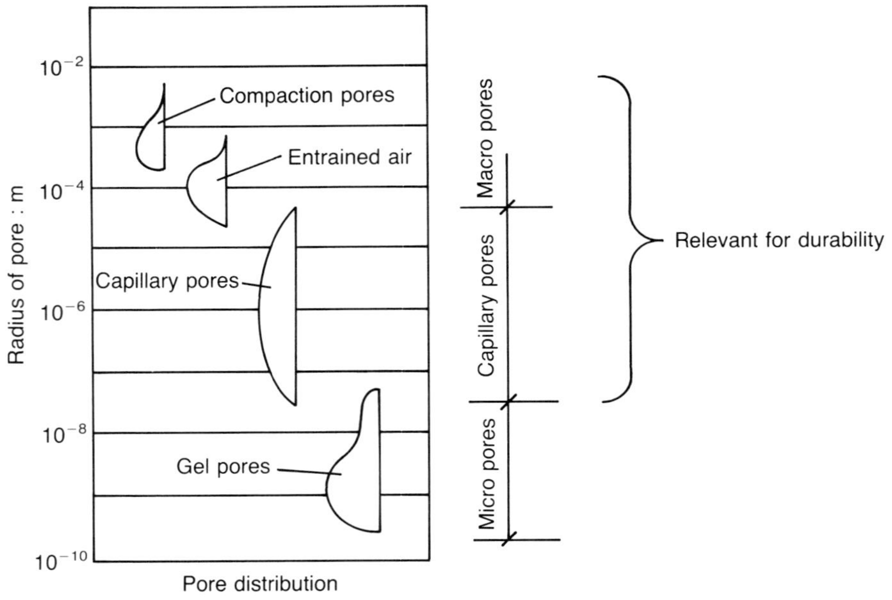

Fig. 2.2. Pore size distribution (according to Setzer)

building materials, two parameters will be of importance: open porosity and pore size distribution. Open porosity means pores which are interconnected so that transport of liquids or gases and/or the exchange of dissolved substances is possible. It corresponds to the maximum reversible water content and, in the case of cement paste, lies in the region of 20–30%.

The pore size distribution particularly influences the rate of the transport. The sizes of pores in the cement paste range over several orders of magnitude. According to origin and characteristics, the pores are described as compaction pores, air pores, capillary pores or gel pores. Expressed in more general terms, it appears to be convenient to classify them as macropores, capillary pores and micropores (Fig. 2.2). The capillary pores and macropores are particularly relevant with regard to durability. In general, the resistance of concrete to chemical and physical influence is considerably reduced with increasing quantity of capillary pores.

2.3. Interaction between pores and water

Free surfaces of solids (e.g. pore surfaces) exhibit a surplus of energy (the surface energy) due to a lack of binding components to the adjacent molecules. In cement paste pores, this surface energy causes the water vapour molecules

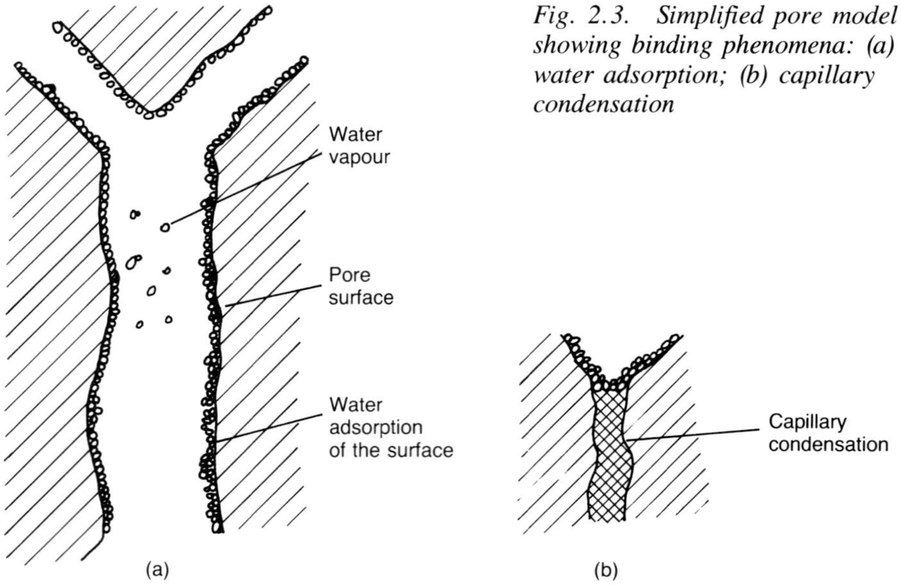

Fig. 2.3. Simplified pore model showing binding phenomena: (a) water adsorption; (b) capillary condensation

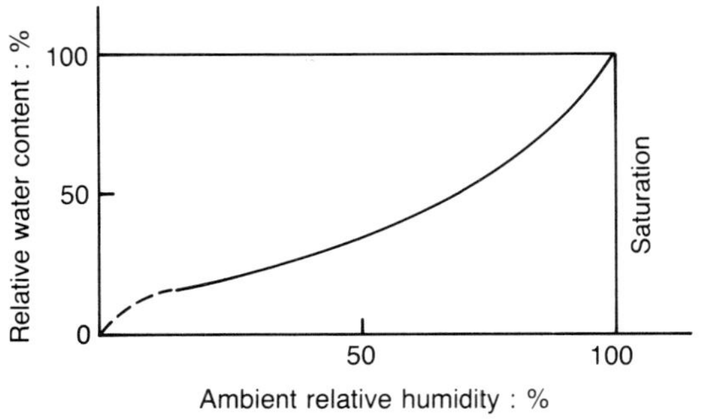

Fig. 2.4. Relationship between relative humidity of ambient air and concrete, relative to saturation

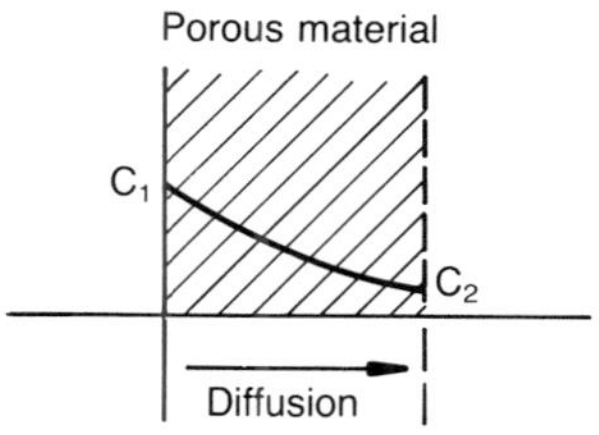

Fig. 2.5 (right). Diffusion through a porous material. The driving force is the difference between c_1 and c_2, where these are the concentrations (or partial pressures or pressures) of water, carbon dioxide, oxygen, chloride ions and so on

within the pores to adsorb onto the pore surface, the thickness of the water film depending on the degree of humidity within the pores (Fig. 2.3).

Due to the fact that the ratio between surface area and volume of the pores increases with decreasing pore radius, the quantity of water adsorbed relative to the pore volume will also increase until, at a certain limit value of the pore radius, the pores with smaller radii are completely filled with water. This process is called capillary condensation. The limit value of the pore radius depends primarily on the water content of the air in the pore which, all else being constant, is proportional to the humidity of the air surrounding the concrete (Fig. 2.4).

As a result of the high proportion and small radii of the gel pores (see section 2.2), concrete exhibits comparatively high water content even when the humidity of the surrounding air is relatively low. Increasing the humidity of the air will cause the larger pores to be filled with water, thus reducing the pore space available for the diffusion of gases. Consequently, the permeability of the concrete with regard to gases will decrease considerably with growing water content and, in the case of an almost water-saturated concrete, the diffusion of gases (e.g. carbon dioxide and oxygen) becomes practically negligible.

2.4. Transport mechanisms in humid air

As outlined above, the larger pores in concrete surrounded by air are filled with air, depending on the humidity of the ambient air. The surface of these pores is coated with a water film bound by adsorption (Fig. 2.3). Any transport processes of gases, water, or substances dissolved in water are diffusion processes under these ambient conditions. Diffusion processes are induced by the tendency for differences in concentrations to equilibrate (Fig. 2.5).

Carbon dioxide diffuses into the concrete due to a chemical reaction between the carbon dioxide and the concrete developing at the pore walls, which causes the concentration within the pores to be reduced. A similar process applies with oxygen when it is consumed during corrosion of the reinforcement.

Diffusion of water or water vapour will always take place when the ambient humidity changes or when the concrete is drying out.

The diffusion of substances dissolved in water (e.g. chloride ions) will develop in the water film at the pore surface or in the water-filled pores. Due to the decreasing film thickness and the decreasing proportion of water-filled pores, respectively, the diffusion rate of substances dissolved in water will be substantially reduced with decreasing moisture content of the concrete.

2.5. Transport mechanisms: rain and splash water

In the case of wetting of concrete surfaces (e.g. rain and splash water), water transport is of major importance (Fig. 2.6). Because of capillary suction, saturation will quickly be achieved. Solutes are transported by the water; the diffusion of gases is practically totally impeded. Only when water transport comes to rest by approaching an equilibrium state does diffusion again play a dominant role.

The effect of capillary suction depends on the surface energy of the pore surface, as described in section 2.3. The tendency to adsorb water onto the

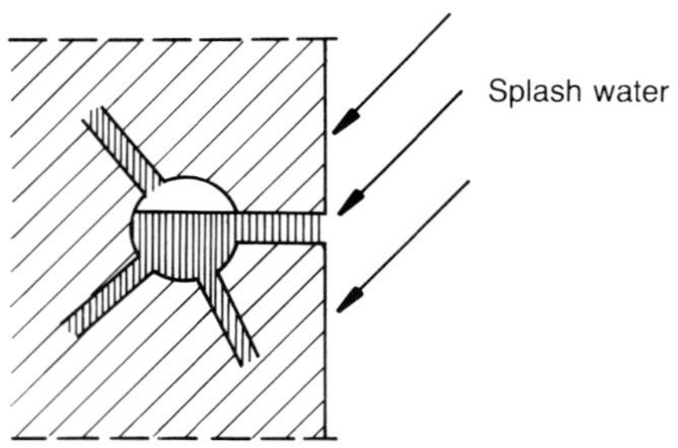

Fig. 2.6. Model of pores in concrete affected by rain

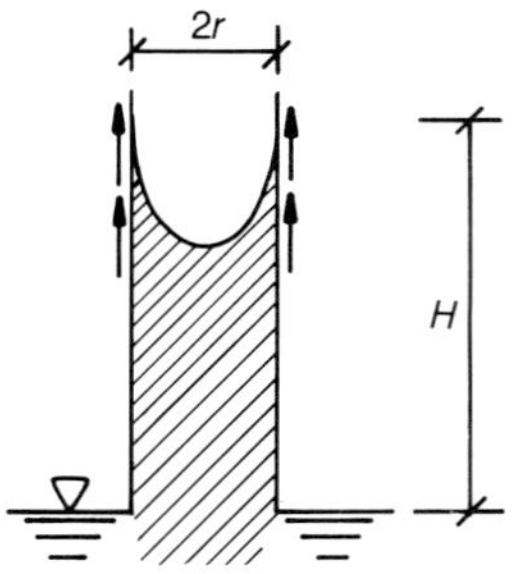

Fig. 2.7 (far right). Capillary suction caused by surface energy. For the vertical capillary shown here, the rise in water level H = *15/r mm, where* r *is the radius of the pore*

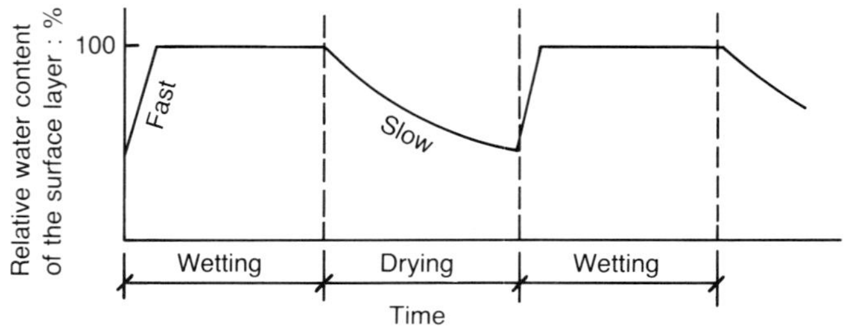

Fig. 2.8. Changing wetting and drying of the surface layer

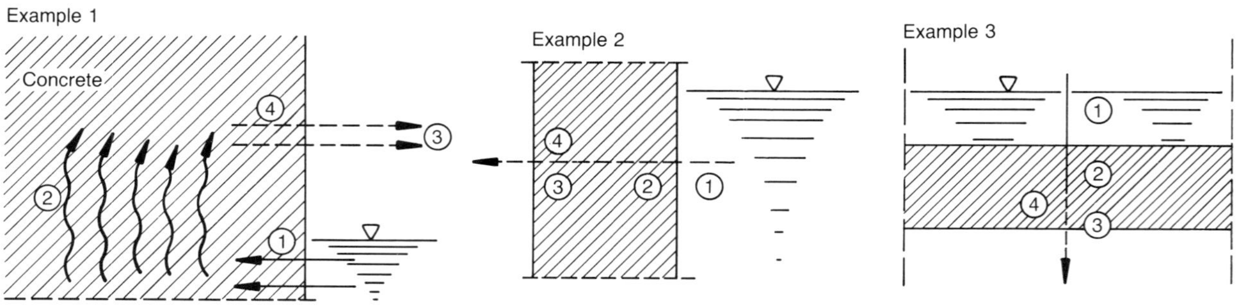

Fig. 2.9. Immersion of concrete in water: 1 = water transport by hydraulic pressure and capillary suction; 2 = transport of water and dissolved agents; 3 = evaporation of water; 4 = crystallization of solutes, giving enrichment in the evaporation zone

surface will, in the case of a surplus of water, result in suction being initiated. The height of capillary rise in vertical capillaries is determined by an equilibrium between the binding forces of the surface and the weight of the water column in the capillary (Fig. 2.7). As far as suction in a horizontal direction is concerned, the depth of penetration will primarily depend on there being an excess of water at the concrete surface and on the duration of this situation.

Water is absorbed by concrete through capillary suction at a considerably higher rate than it is disposed of by evaporation (Fig. 2.8).

2.6. Transport mechanisms: immersion

In the case of continuously immersed structures large quantities of water may, under unfavourable conditions, be transported. The penetration of water will first take place by capillary suction, possibly accelerated by an increased hydraulic pressure. Continuous transport of water will develop only when water is allowed to evaporate at the concrete surfaces exposed to the air. The intensity of this water transport depends on the relationship between evaporation, capillary suction and hydraulic pressure (Fig. 2.9).

Along with the water, dissolved agents (e.g. carbonates, chlorides and sulphates) will be transported. However, these agents are left behind in the concrete in the evaporation region where they are likely to develop considerable concentrations. Efflorescence phenomena may be due to this effect: the dissolved agents recrystallize at the concrete surfaces.

In concrete the expansive forces due to salt crystallization near the surface cause only minor problems; of more importance is the chemical effect of the increased concentration of aggressive substances. However, in other porous materials such as sandstone, marble or masonry bursting and scaling due to salt crystallization is a serious cause of deterioration and results in rapid deterioration of sculptures, monuments, etc. exposed to aggressive environments.

3. Physical processes in concrete

3.1. Cracking

3.1.1. Causes of cracking

Cracking will occur whenever the tensile strain to which concrete is subjected exceeds the tensile strain capacity of the concrete. The tensile strain capacity of concrete varies with age and with rate of application of strain.

There are various basic mechanisms by which strains may be generated.

(*a*) Movements generated within the concrete. Examples are drying shrinkage, expansion or contraction due to temperature change, and plastic settlement or shrinkage. These effects only cause tensile stresses if the movements are restrained. This restraint may be local, for example where the shrinkage of concrete is restrained by the reinforcement, or on a larger scale, as for example where a member is restrained from shrinkage by the members to which it is connected.

(*b*) Expansion of material embedded within the concrete. An example of this is corrosion of reinforcement.

(*c*) Externally imposed conditions. Examples of these are loading or deformations imposed by differential settlement of foundations.

Figure 3.1 summarizes various possible causes of cracking, and Fig. 3.2 gives some indication of the age at which the various forms of cracking can be expected to occur.

Mechanisms (*a*) and (*b*) cause various types of intrinsic crack, for which more details are given in Fig. 3.3 and Table 3.1. Mechanism (*c*) causes extrinsic cracks. The types of cracks which occur most often in practice are described below.

'Young' concrete is especially prone to cracking (Fig. 3.4). During the transition phase leading from fresh ('green') to hardening ('young') concrete, a critical period with low tensile strength and a low deformability starts a few hours — at the earliest 2 h — after casting and lasts about 4–16 h (Fig. 3.5).

3.1.1.1. Plastic shrinkage and plastic settlement cracking. There are two distinct types of plastic cracking: plastic shrinkage cracking, which most commonly occurs in slabs, and plastic settlement or slump cracking, which

Fig. 3.1. Types of crack[5]

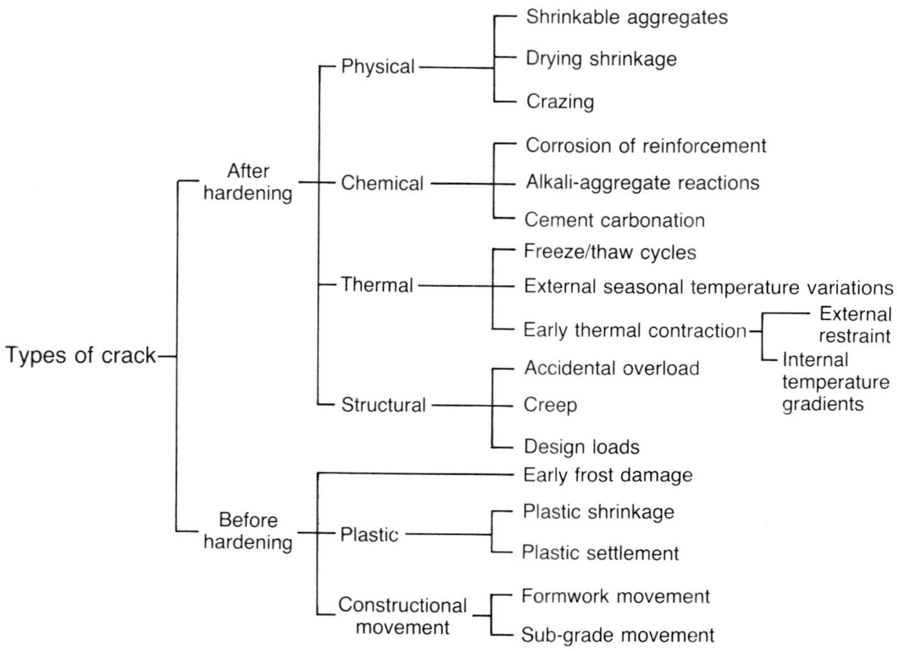

Fig. 3.2. Time of appearance of cracks from placing of concrete

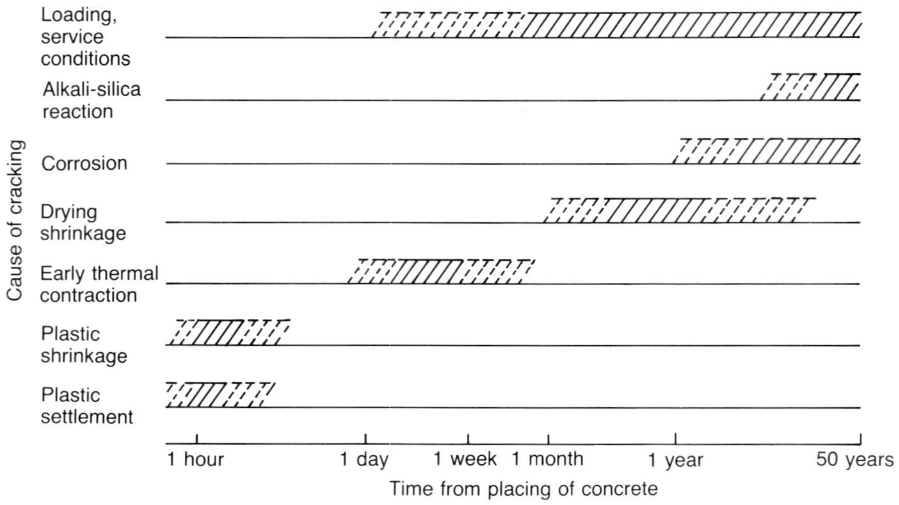

Table 3.1. Classification of intrinsic cracks[5]

Type of cracking	Position on Fig. 3.3	Subdivision	Most common location	Primary cause (excluding restraint)	Secondary causes/ factors	Remedy (assuming basic redesign is impossible). In all cases reduce restraint	For further details see section . . .	Time of appearance
Plastic settlement	A	Over reinforcement	Deep sections	Excess bleeding	Rapid early drying conditions	Reduce bleeding (air entrainment) or revibrate	3.1	10 minutes to 3 hours
	B	Arching	Top of columns					
	C	Change of depth	Trough and waffle slabs					
Plastic shrinkage	D	Diagonal	Roads and slabs	Rapid early drying	Low rate of bleeding	Improve early curing	3.1	30 minutes to 6 hours
	E	Random	Reinforced concrete slabs					
	F	Over reinforcement	Reinforced concrete slabs	Rapid early drying, steel near surface				
Early thermal contraction	G	External restraint	Thick walls	Excess heat generation	Rapid cooling	Reduce heat and/or insulate	3.1 Appendix 1	1 day to 2–3 weeks
	H	Internal restraint	Thick slabs	Excess temperature gradients				
Long-term drying shrinkage	I		Thin slabs (and walls)	Inefficient joints	Excess shrinkage, inefficient curing	Reduce water content, improve curing	3.1	Several weeks or months
Crazing	J	Against formwork	'Fair-faced' concrete	Impermeable formwork	Rich mixes	Improve curing and finishing	3.1	1–7 days, sometimes much later
	K	Floated concrete	Slabs	Over-trowelling	Poor curing			
Corrosion of reinforcement	L	Natural	Columns and beams	Lack of cover	Poor quality concrete	Eliminate causes listed	6.2	More than 2 years
	M	Calcium chloride	Precast concrete	Excess calcium chloride				
Alkali-aggregate reaction	N		(Damp locations)	Reactive aggregate plus high-alkali cement		Eliminate causes listed	4.4	More than 5 years

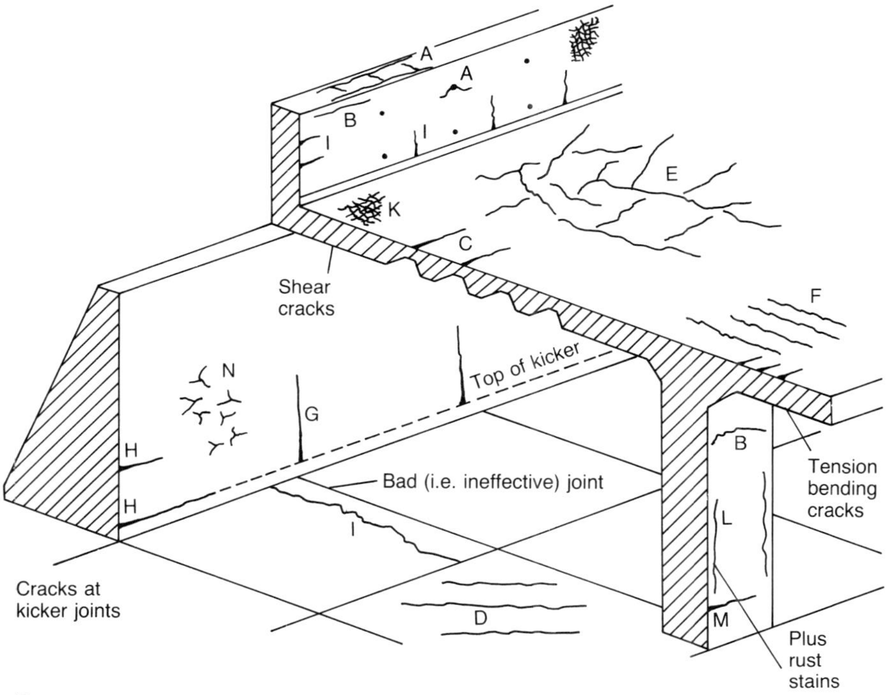

Fig. 3.3. Examples of intrinsic cracks in a hypothetical concrete[5] (letters refer to Table 3.1)

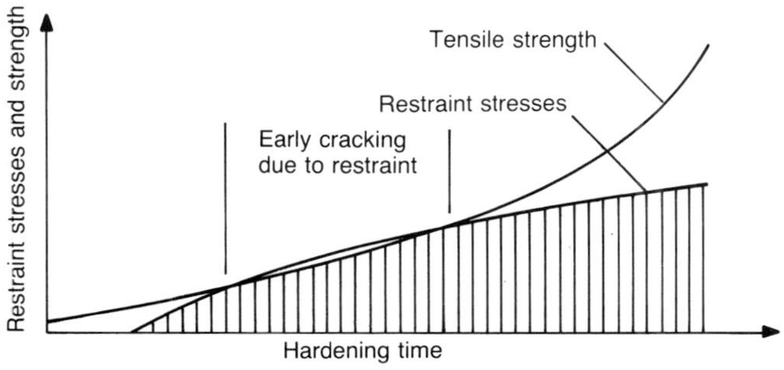

Fig. 3.4. Evaluation of strength and restraint stresses in young concrete

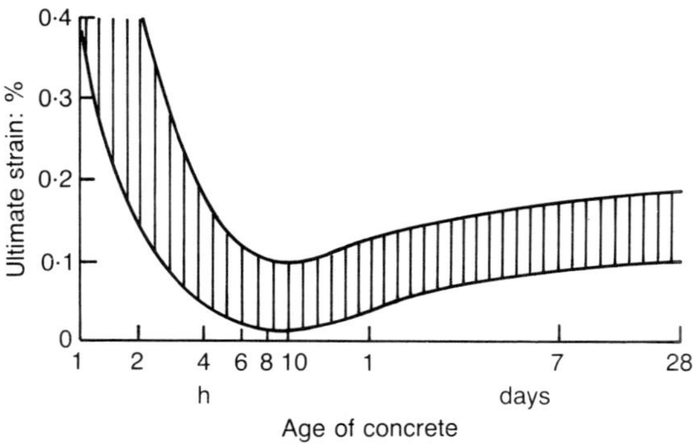

Fig. 3.5. Ultimate tensile strain of concrete as a function of age

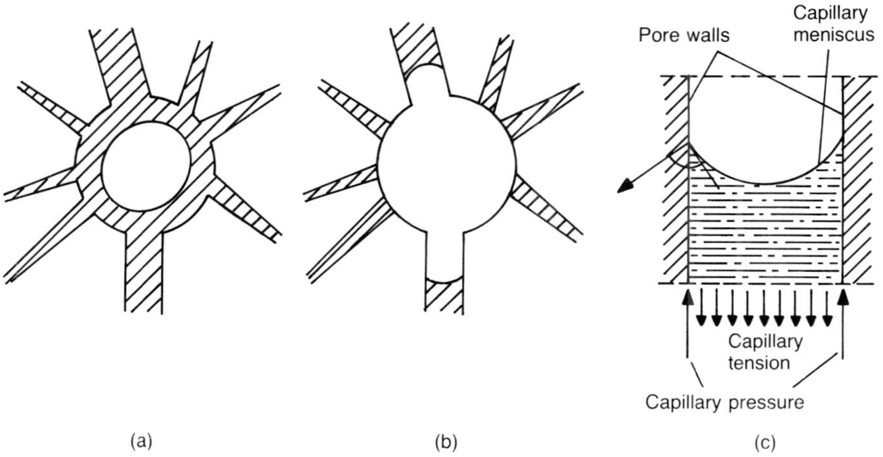

Fig. 3.6. Behaviour of water in narrow pores: (a) saturation; (b) drying out; (c) capillary pressure

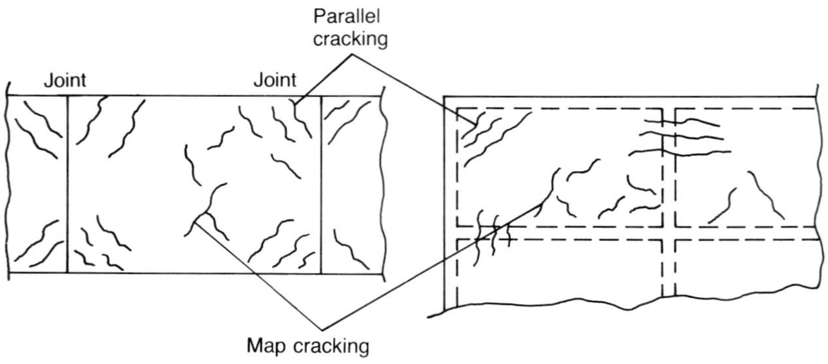

Fig. 3.7. Plastic shrinkage cracks in the surface of concrete pavements and continuous floor slabs

may occur in deep members. Both types are associated with bleeding of the concrete.

Plastic shrinkage is a characteristic property of 'green' concrete. It is caused by capillary tension in the pore water. Plastic shrinkage cracking occurs within the first 2–4 h after mixing, shortly after the disappearance of the wet shine when the concrete surface becomes mat, if the loss by vapourization exceeds the supply by bleeding water, thereby activating capillary forces in the pore water (Fig. 3.6). If the volume decrease is hampered in zones near the surface (e.g. by coarse aggregate below the surface or the reinforcement) the probability of cracking is high because the tensile stress is not countered by any tensile strength.

Concrete parts with extended horizontal surfaces, such as slabs, are prone to cracks by plastic shrinkage. Parallel cracks in slabs at an angle of about 45° to the slab corners are typical; the crack spacings are irregular and fall in the range 0·2–1 m (Fig. 3.7). Figure 3.7 also shows another typical kind of cracking, known as map cracking.

Cracks caused by plastic shrinkage are mostly surface cracks, but in a few cases they can penetrate a whole slab, the crack width decreasing considerably with increasing depth from the surface. Typical crack widths are of the order of 2–3 mm at the surface.

Fig. 3.8. Longitudinal crack: settlement crack along bar

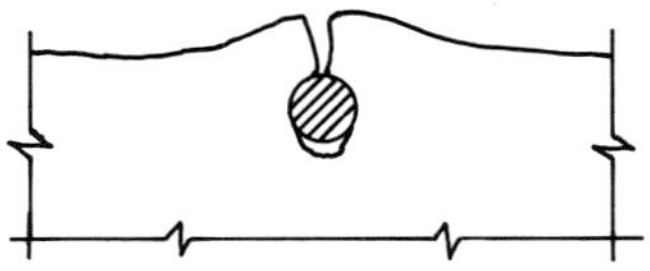

During settlement, the concrete bleeds. As a result of gravitational forces, the concrete particles settle and the displaced mixing water surfaces. Due to this decrease in volume, the concrete settles in the formwork. If settlement of concrete is hampered by the reinforcement or by the formwork, cracking can occur. Such cracks are longitudinal (Fig. 3.8), following the direction of the reinforcement on the top of deep beams (Fig. 3.9(a)) or thick slabs, or the stirrups at the lateral surfaces of columns (Fig. 3.9(b)).

Of special concern is the horizontal settlement cracking which may occur

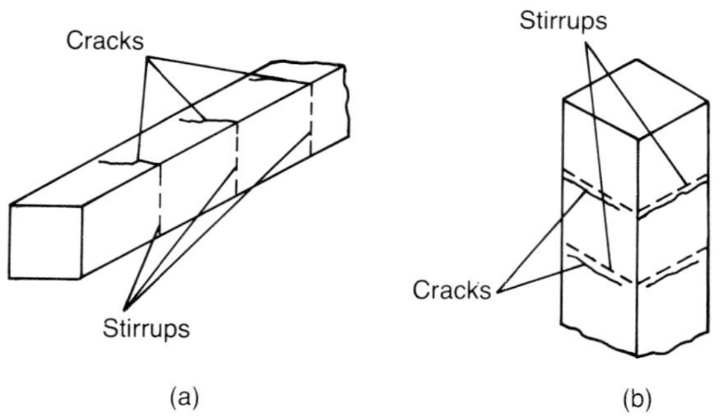

Fig. 3.9. Cracks due to plastic settlement: (a) in the direction of reinforcement on the top of a deep beam; (b) at the stirrups at the lateral surfaces of a column

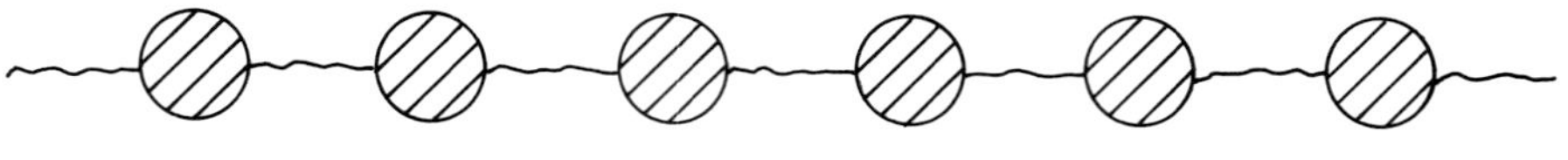

Fig. 3.10. Horizontal settlement cracking between closely spaced reinforcing bars

when the reinforcing bars are closely spaced (Fig. 3.10). These cracks cause delamination of the concrete cover on the top layer of the reinforcement. In unfavourable situations the bottom cover may also delaminate, creating the risk of unexpected spalling of the concrete cover. When this is followed by deterioration mechanisms of an expansive nature, such as frost or reinforcement corrosion, there is a danger of a sudden unpredictable spalling of major parts of the concrete cover, endangering the users of the structure.

3.1.1.2. Cracking caused by direct loading. Cracking caused by direct loading covers cracking resulting from normal load effects (i.e. bending, shear, tension, etc.) applied to sections. The following points should be noted.

(*a*) In any section containing bonded reinforcement arranged more or less perpendicularly to the expected direction of the principal tensile stress with covers in accordance with the model code,[6] cracking is likely to be relatively small ($<0\cdot5$ mm) under service loads. This will be true even where no direct action is taken to control the cracking, provided that the reinforcement does not yield under the service load.

(*b*) Although in laboratory tests large numbers of fairly closely-spaced cracks may be obtained, this is not generally the case in practice, since actual service loads are rarely high enough to generate anything approaching the 'final' crack pattern obtainable in laboratory tests. A few cracks at points of maximum stress are the most that are normally found.

(*c*) Where wide load-induced cracks are found, they are almost always an indication that the calculations for the ultimate limit state are incorrect. This may be due to mistakes or to the effects of a particular form of loading being misunderstood or neglected to the extent that no or insufficient reinforcement has been provided to resist a particular load effect.

Cracking may also result from overstressing the concrete locally. Common examples are cracking due to excessive bond stresses leading to cracking along the line of the bar, and cracking due to concentrated loads such as those beneath anchorages of prestressing tendons leading to cracking parallel to

Fig. 3.11. Load-induced cracks: (a) pure flexure; (b) pure tension; (c) shear; (d) torsion; (e) bond; (f) concentrated load

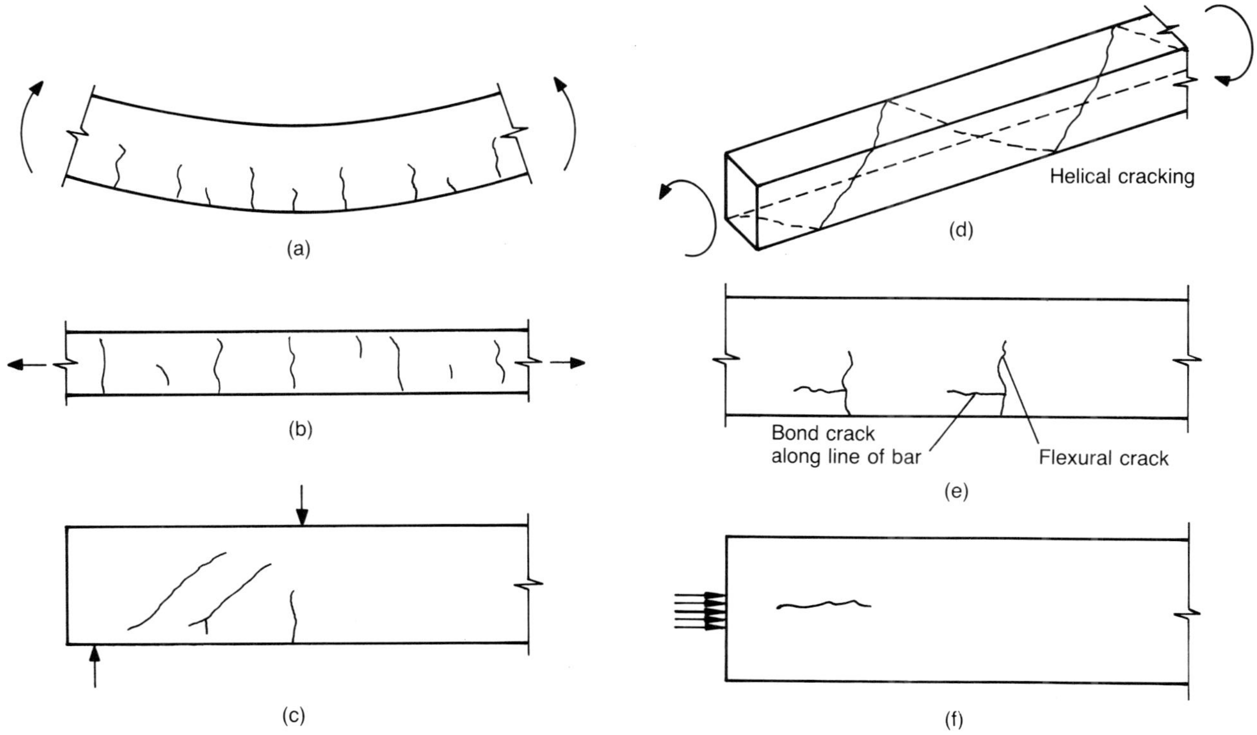

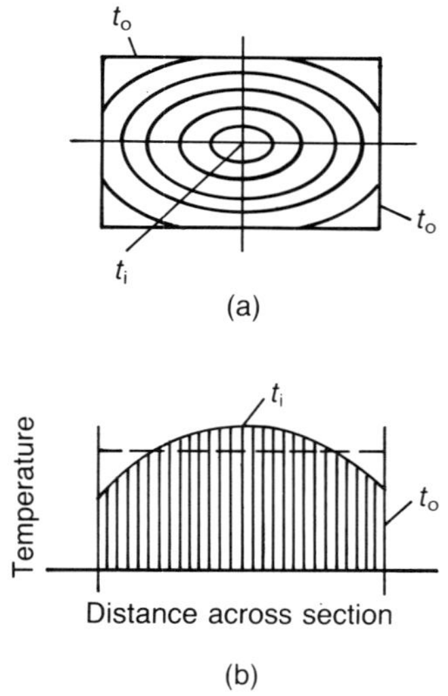

Fig. 3.12. Distribution of temperature due to hydration heating: (a) cross-section, showing lines of equal temperature; (b) mid-span section

Fig. 3.13. (a) Stresses due to temperature (self-equilibrating stresses); (b) map cracking due to the self-equilibrating stresses

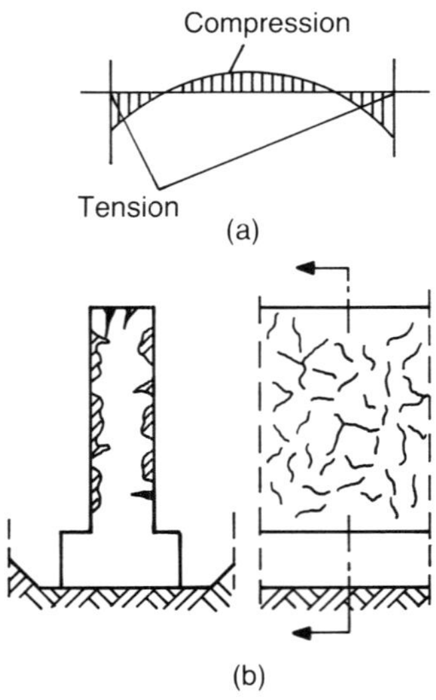

the direction of the applied compression, usually starting some way from the surface where the loading is applied.

Figure 3.11 summarizes the various forms of load-induced cracking which may occur and shows their general form.

3.1.1.3. Cracking resulting from imposed deformations. This section considers cracking resulting from causes such as temperature, shrinkage or differential settlement of foundations. The common feature of these is that stresses, and hence cracking, can arise where the structure, or a member or part of a section, resists the imposed movement. The greater the degree of restraint provided by the structure, the higher will be the stresses, and the larger will be the cracks.

Temperature differences are frequent causes of cracking. One of the major causes of cracking in structures is movement resulting from the cooling of members from the heat generated by hydration of cement. Cracking due to early thermal movements was once commonly diagnosed as shrinkage cracking.

The hydration heat of cement, which is set free during the setting and hardening of concrete, cannot be passed on rapidly enough to the surrounding air by the concrete surface, especially in the case of massive parts. A temperature gradient from the core to the surface of the concrete part develops, which increases with increasing temperature of concrete and decreasing air temperature (Fig. 3.12). A condition of self-equilibrating stresses is created, with tensile stresses in the outer layers and compressive stresses in the core. If the tensile stresses exceed the still low tensile strength of hardening concrete, cracks are formed (Fig. 3.13).

The cracks are always surface cracks, mostly in the form of map cracking. They are normally a few millimetres or centimetres in depth and usually close up when temperature differences vanish. However, they become visible again when the surface is wetted (e.g. by rain) and then dries up again; the moisture sucked into the cracks reveals their permanent existence.

In the normal case of unequal, non-linear temperature distribution, a structural element is changed in length and bent. If these deformations are restricted, restraint stresses develop, which are superimposed on the self-equilibrating stresses caused by the non-linear temperature distribution.

If a structural element is stressed, especially by axial or eccentric tension, partition cracks are formed which penetrate the whole cross-section of the element. Figure 3.14 shows a typical starting point for the formation of such partition cracks, when rising walls of greater sections, e.g. for cellars, tank constructions or abutments, are placed on already hardened foundations.

Stresses caused by differential shrinkage develop gradually with the long-term drying of the concrete, whereby the simultaneous effect of creep reduces the resulting stresses. This favourable effect of creep is not encountered in the development of stresses due to differential temperature caused by heat of hydration, since this process takes place up to a few days after casting, thus involving a young concrete with low deformability.

Cracking can be caused in structures in service by temperature differences within members. A chimney, for example, which can be hot on the inside and relatively cool on the outside, can develop vertical cracks on the outside. Sudden cooling, for example during the emergency shutdown of a reactor pressure vessel, can also lead to serious cracking. Due to diurnal variations in the environment, markedly non-linear temperature distributions can be set up within, for example, the deck structure of a bridge or pavement. These can induce stresses sufficient to cause cracking which, if not controlled by the presence of adequate reinforcement or prestress, can be unacceptable.

Shrinkage is the load independent, long-term deformation of concrete because of its decrease in volume due to drying. If the shortening of a structural element due to shrinkage is restrained from the outside, axial or eccentric forces develop, producing separation cracks if the ultimate strain of concrete is exceeded.

When concrete dries out from the surface, differential shrinkage between the surface layer and the core causes a state of equilibrating stresses to develop with tensile stresses at the surface and compressive stresses in the core. Like cracking due to temperature, surface cracking caused by shrinkage is mostly map cracking and is frequently undistinguishable from cracking due to temperature (Fig. 3.13).

Shrinkage is at least partially reversible and, where there is an increase in humidity, significant swelling can occur. Shrinkage movements are not confined only to the early life of the structure. A drop in relative humidity (possibly due to change in central heating or air-conditioning procedures) at any time during the life of a structure can be the cause of significant movements and crack development.

Cracking due to settlement of foundations mainly affects non-structural elements, such as partitions, infill panels, windows and doors, unless the differential settlements are substantial. In the latter case, cracks similar to load-induced cracks may develop.

3.1.1.4. Alignment of cracks relative to the reinforcement. The importance of cracking relative to the durability and service life performance of a structure may be critically influenced by whether or not cracks are longitudinal, i.e. follow the line of the reinforcing bars (Fig. 3.15).

This is especially important from the point of view of reinforcement corrosion, as discussed in section 6.2.6, but in addition bond and shear strength could be seriously reduced by the development of longitudinal cracks.

Cracking caused by tension or bending under direct loading or imposed deformations will be expected to form perpendicularly to the direction of the main reinforcing bars being placed in the direction of principal tension. Such loading is unlikely to cause cracking longitudinal to the main bars. However, there will commonly be some transverse reinforcement present, and frequently such cracks will form along the line of the transverse bars; indeed, such bars may act as crack initiators (Fig. 3.15).

Shear and tension lead to diagonal cracks which are unlikely to coincide with the line of the reinforcing bars.

Bond cracks will form along the line of the main bars, but in an appropriately designed structure these cracks are unlikely to occur under service loads to any significant extent.

Plastic shrinkage cracks may, by chance, follow the line of the reinforcement. Clearly, this is true of plastic settlement cracks (slump), where the

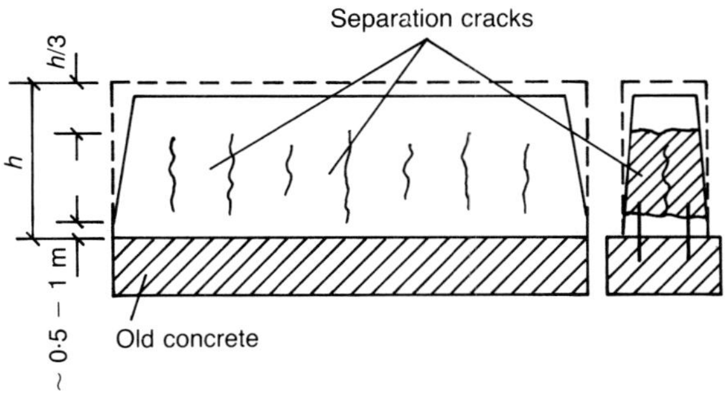

Fig. 3.14. Cracking due to early thermal movements in a wall

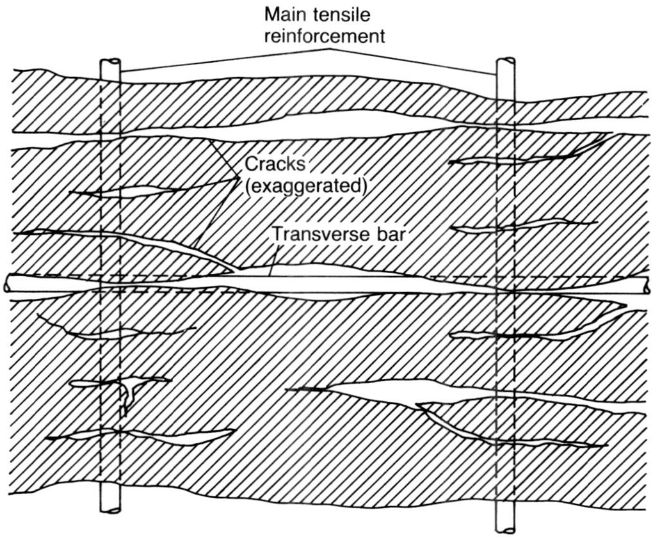

Fig. 3.15. Alignment of cracks relative to reinforcement

cracks are often directly caused by the bars. The risks of obtaining cracking along the lines of some reinforcing bars are high; transverse reinforcement is particularly at risk, especially in cases where it has a lower concrete cover than the main bars, such as stirrups in beams.

3.1.2. Influencing parameters

Although numerous crack prediction formulae have been proposed to cope with load-induced cracking, it should be noted that the crack prediction formula given in the model code[6] has probably been tested against a larger body of data than most other formulae.

All the formulae considered deal with cracking caused by bending or tension, or bending and axial load. The prediction of crack widths caused by shear, torsion or other forms of loading has been much less exhaustively studied.[7]

It is commonly agreed that the types of cracks induced by loads or imposed deformations occurring under normal use do not have serious detrimental effects, provided that the structure is otherwise sound (see, e.g., section 6.2.7).

The more important parameters which determine whether cracking is detrimental to concrete structures are related to the detailing of the structural form and of the reinforcement, to the selection of concrete composition, and to the type and quality of execution and curing.

3.1.2.1. Structural detailing. Abrupt changes of geometry such as depth or cross-sectional area cause differential plastic settlement leading to cracking, or induce local stress concentrations which sooner or later may create cracks. Examples are ribbed slabs, trough sections, waffle slabs or voided slabs.

The number and size of cracks caused by imposed deformations depend on the degree of restraint, external or internal. Internal restraints, e.g. between thin and thick parts of the section or between the core and the surface layer of a section, are influenced by the maximum temperature differences occurring during initial hardening and during ordinary use, and by the selected detailing of the corresponding reinforcement.

3.1.2.2. Detailing of reinforcement. Reinforcement may initiate cracks either where concentrated forces are transmitted to the concrete or where the reinforcement unfavourably influences the placing and setting of the concrete.

Concentrated forces occur at sharp bends, at curtailed reinforcement, at laps, in zones with high bond stresses, near anchorages for prestressing tendons and so on.

In the detailing of the reinforcement, the actual concrete cover and the bar spacings are decisive factors in assuring appropriate placing and compaction of concrete, especially in heavily reinforced zones such as those near supports or at intersections of beam, column or slab elements.

3.1.2.3. Concrete composition. The composition of concrete mainly influences the plastic shrinkage and settlement cracking, which depends on the bleeding of the concrete. Bleeding can be diminished and even avoided altogether by carefully selecting the grading of the aggregates, choosing a blended cement, and using plasticizing or superplasticizing admixtures. Hence the risk of settlement or slump cracking is reduced, but at the same time the risk of plastic shrinkage cracking is increased.

3.1.2.4. Execution and curing. The workmanship associated with the execution process has a decisive influence on the homogeneity and uniformity of cast concrete as well as on the correct placement of the reinforcement. The concrete cover to the reinforcement and the quality (i.e. low permeability) of the outer surface layer of the concrete (the skin) are basic parameters influencing the subsequent resistance of the whole structure to an aggressive environment.

Cracking developed during the execution process and during the initial period of hardening may be the main initial cause for a later acceleration of deleterious actions which depend on water or aggressive substances (e.g.

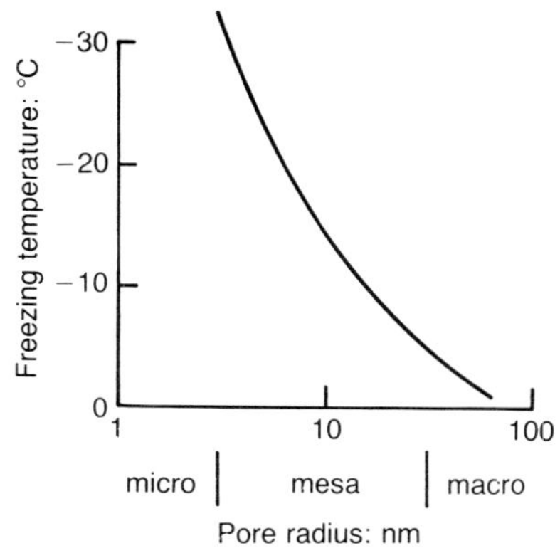

Fig. 3.16. Depression of freezing point due to surface energy

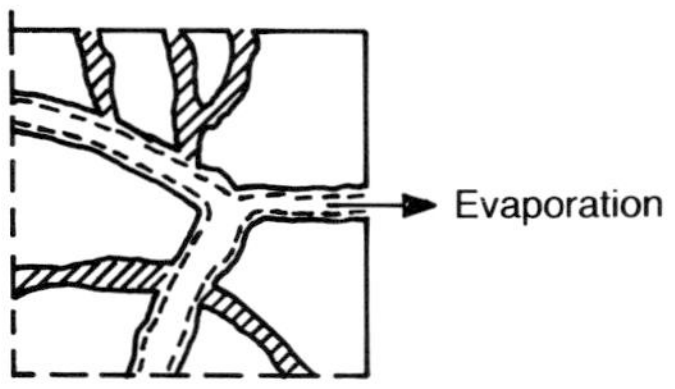

Fig. 3.17. Evaporation during cooling

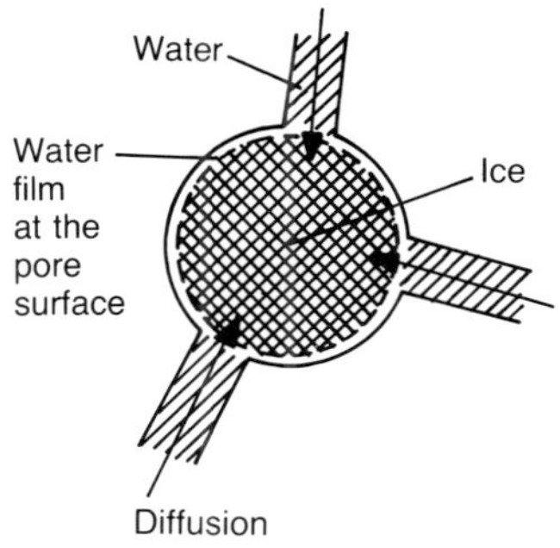

Fig. 3.18. Diffusion during cooling

carbon dioxide, acids and sulphates) entering from the outside through the outer concrete layer.

3.2. Frost and de-icing agents

3.2.1. Deterioration mechanisms

In the case of water freezing in porous building materials, such as cement paste, four physical processes are of major importance, as they determine the freezing resistance by their mutual interaction and, in particular, the resistance of the concrete to freezing and thawing cycles.

Transition from water to ice involves an increase in volume by 9%. In the case of completely water-filled pores, this will cause splitting of concrete.

The surplus energy at the pore surface results in a reduction of the potential energy of the pore water and, thus, in a depression of the freezing point. Due to the wide range of pore radii of cement paste, only about one third of the pore water will be frozen at a temperature of −30°C (−22°F) and only two thirds will be frozen at −60°C (−76°F). A thin film of water coating the pore surfaces will remain even after the pore water has formed ice (Fig. 3.16).

Transition from water to ice in porous systems is likely to cause a relatively large quantity of water to evaporate, if ambient conditions (e.g. air) and the degree of saturation of the concrete allows (this will not occur in completely water-saturated concrete) (Fig. 3.17).

Another consequence of the surface energy is a hydraulic underpressure that develops in the smaller pores during cooling, inducing the diffusion of water not yet frozen from the smaller pores to the larger ones in the concrete (Fig. 3.18).

3.2.1.1. Critical saturation and the effect of air entrainment. Owing to the fact that the volume of water increases during freezing and diffusion also takes place during cooling, a sufficient quantity of pores not filled with water should be available to allow the water to expand, thus preventing damage by frost. The limit value of the water content causing damage to occur is defined by the critical degree of saturation. This depends primarily on

(*a*) the age of the concrete (which determines the degree of hydration and pore structure)
(*b*) pore size distribution (including artificial air pores)
(*c*) environmental conditions (i.e. how easy it is for the water to evaporate)
(*d*) the rate of cooling and frequency of freezing and thawing cycles (redistribution of water)
(*e*) drying out between freezing and thawing cycles (provision of additional expansion space).

Artificial air pores may be defined as quasi-closed pores. They are not

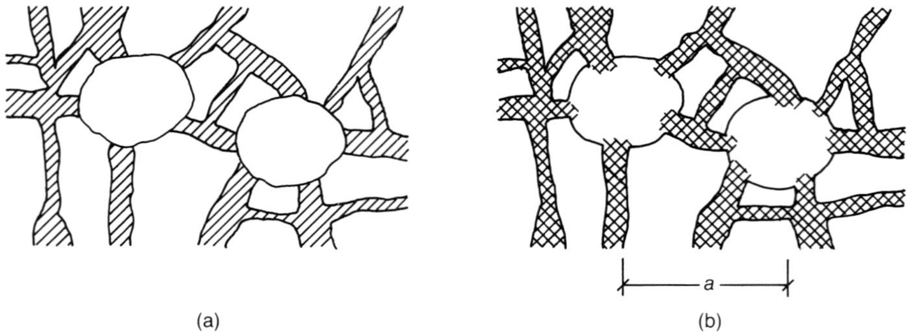

Fig. 3.19. Effect of air entrainment: (a) artificial air pores, not filled with water even in the case of water saturation; (b) air pores provide expansion space for freezing water

filled with water even in the case of saturated concrete. However, by diffusion processes during freezing of water they may well be reached by the water forming ice and are thus available as expansion space (Fig. 3.19). Their spacing a must not exceed a particular maximum value so as to ensure their efficiency in the pore system. The critical spacing a_{crit} will be lower with increasing severity of the frost attack.

As the diffusion processes during freezing of the water are to some extent irreversible, the filling up of the larger pores with water will increase as the number of freezing and thawing cycles increases. This means that in certain circumstances damage by frost will occur only after a series of freezing and thawing cycles, provided that there is no possibility of (at least partial) drying of the concrete between the individual cycles.

3.2.1.2. Effect of de-icing agents. The application of de-icing agents to a concrete surface covered with ice will cause a substantial drop in temperature at the concrete surface (temperature shock) during thawing of the ice. The difference in temperature between the surface area and the interior of the concrete gives rise to a state of internal stresses likely to induce cracking in the region of the outer layer of the concrete (Fig. 3.20).

Another significant effect is a change in the freezing behaviour of the pore water due to de-icing agents penetrating from the outside of the concrete (Fig. 3.21). As explained, the freezing point of the pore water will be lower when the pore radius is smaller. The diffusion processes in the pore water will further cause the content of de-icing agents in the pore water to be reduced with decreasing radius. This will lead to a less noticeable dependence of the freezing point on the pore radius. Moreover, the content of de-icing agents

Fig. 3.20. Distribution of tensile strain in concrete experiencing thermal shock at the surface due to the effects of chlorides

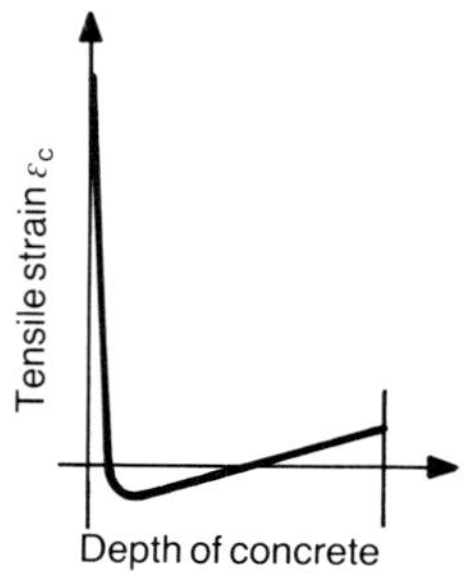

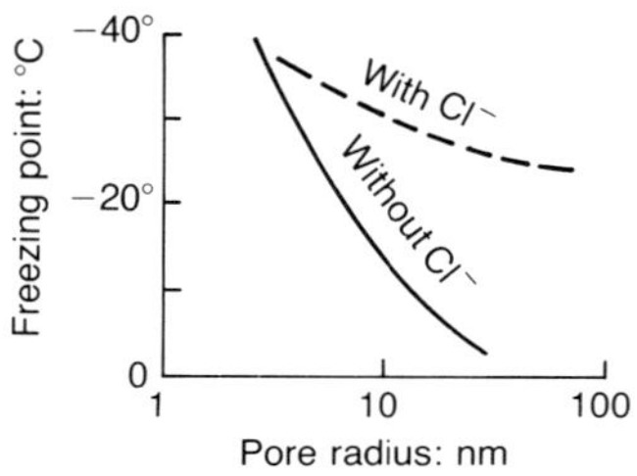

Fig. 3.21. Effect of chlorides on the freezing properties of pore water

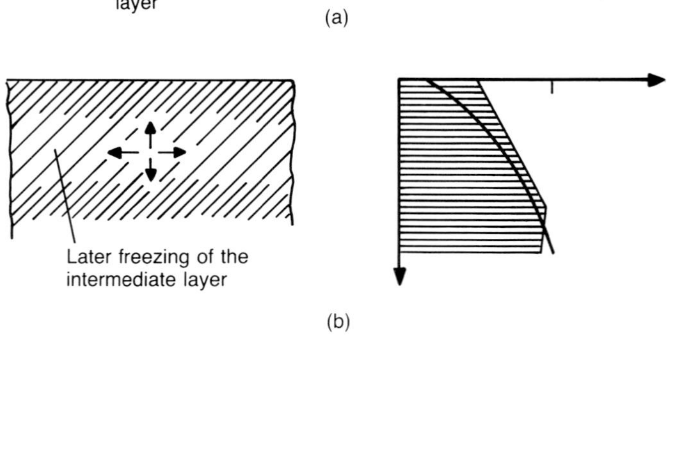

Fig. 3.22. Scaling due to variations in the timing of freezing of layers: (a) intermediate layer is initially unfrozen; (b) intermediate layer freezes later, causing scaling

Fig. 3.23. Pop-out due to non-frost-resistant aggregates

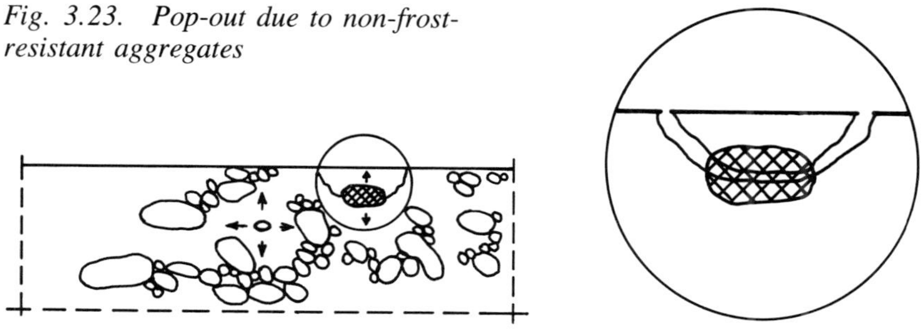

will decrease with increasing distance from the surface of the concrete.

The result of both effects is that in the region of larger pores, as well as at greater depths, water freezes within a smaller temperature range, which causes the redistribution of water to be considerably reduced.

As a consequence both of the change in temperature and of the change in content of de-icing agents with increasing distance from the concrete surface, it may happen that certain concrete layers suffer freezing at different times (Fig. 3.22). In this case, scaling may result.

For the reasons outlined above, any frost attack should be considered to be more severe in the presence of de-icing agents. Consequently, to ensure frost resistance under these circumstances a higher content of air pores will be required.

The principles described hold good for all de-icing agents. In the case of chlorides, the de-icing salts most frequently applied, the serious risk of corrosion developing at the reinforcement has to be considered (see sections 6.2.3 and 6.2.4). When using other de-icing agents, the possibility of an additional chemical attack must be taken into account.

3.2.1.3. Influence of aggregates. Aggregates which are not frost-resistant will, as a rule, absorb water that expands during freezing and destroys the cement paste. Typical indications of such processes are local spallings above larger-sized aggregates (pop-outs) (Fig. 3.23).

3.2.2. Influencing parameters

3.2.2.1. Concrete composition. The intrinsic influencing factor with regard to frost resistance is the presence of a certain quantity of air pores, which should be adapted to the environmental conditions. The frost resistance of the concrete can thereby be substantially improved; in the case of a severe frost attack, air entrainment can reduce the relative weight loss to 10–20% of that of concrete without air entrainment.

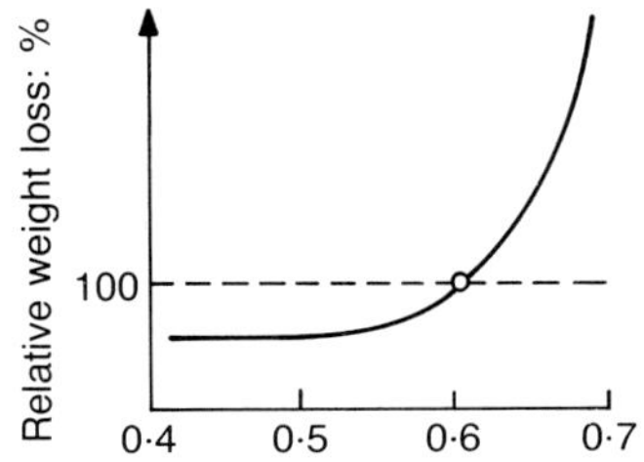

Fig. 3.24. Effect of W/C ratio on relative weight loss during a severe frost attack (cyclic frost–thaw action)

Some further significant parameters are the water/cement (W/C) ratio and the cement content. With the W/C ratio decreasing and the cement content increasing, the frost resistance of the concrete will clearly increase (Fig. 3.24).

A growing content of blending agents will cause a change in the pore structure. High proportions of blending agents may influence the scaling resistance of the concrete unfavourably.

The particle size distribution also influences the frost resistance. With a decrease in the proportion of larger aggregates, an increase in cement and air content will be required to arrive at a frost resistance of equal strength.

3.2.2.2. Environmental conditions. Ambient conditions are the governing criterion with regard to the frost resistance of concrete. Even slight drying out of the concrete before freezing will ensure extremely high frost resistance independent of the W/C ratio and the air content. Ambient moisture conditions showing a relative humidity of approximately 97% will make possible such

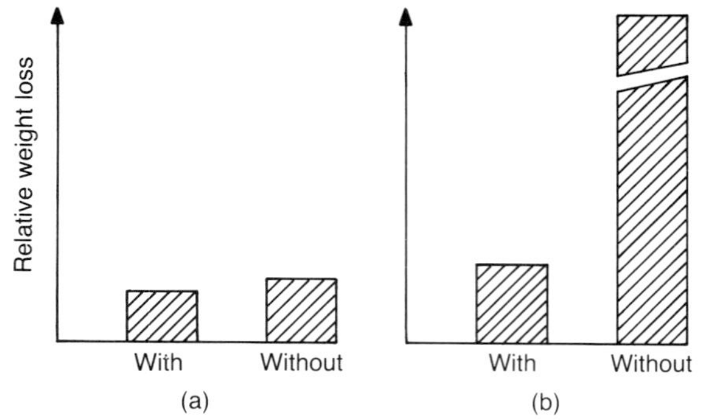

Fig. 3.25. Relative weight loss of concrete with and without previous drying out, during a severe frost attack: (a) 97% relative humidity; (b) in saturated condition

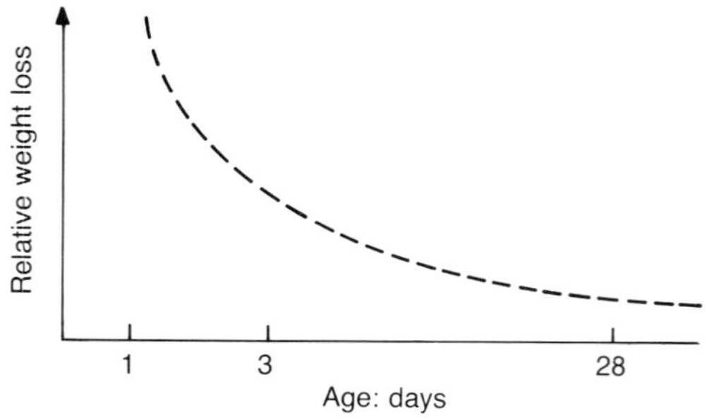

Fig. 3.26 (above right). Effect of age of concrete on relative weight loss during severe frost attack

a high degree of evaporation during the freezing of water that sufficient space will be available for the volume to increase and for the redistribution of water (Fig. 3.25).

It is only in the case of nearly saturated concrete that the influences of concrete composition, illustrated in the preceding section, will have a significant bearing on the frost resistance of the concrete.

3.2.2.3. Age of concrete. As a result of the increasing strength of the concrete and the changing pore structure, frost resistance grows stronger as the age of the concrete increases (Fig. 3.26).

Furthermore, it should be noted that even in ambient humidities not likely to cause damage by frost, concrete at a very early age shows a high moisture content, and thus a confined expansion space. This is due to the fact that the surplus water from the manufacturing process has not yet been disposed of.

3.3. Erosion

3.3.1. Deterioration mechanisms

3.3.1.1. Erosion by abrasion. Abrasive wear of the concrete surface can be caused, for example, by the grinding action of pedestrian traffic on floors, by the scraping, percussive impact of studded tyres on pavements or by impact or sliding of loose bulk materials (Figs 3.27 and 3.28). Abrasive wear can also be caused by the action of heavy particles suspended in water, especially at high water velocities. Such wear occurs, for example, at dams or hydroplants, at constructions for stream regulation, at structures protecting embankments or coasts and at bridge piers.

3.3.1.2. Erosion by cavitation. If water without solids is flowing rapidly parallel to a limiting surface, any change in geometry of the surface causes a flow detachment and zones of low pressure at the limiting surface. If the static pressure of streaming water becomes lower than the vapour pressure

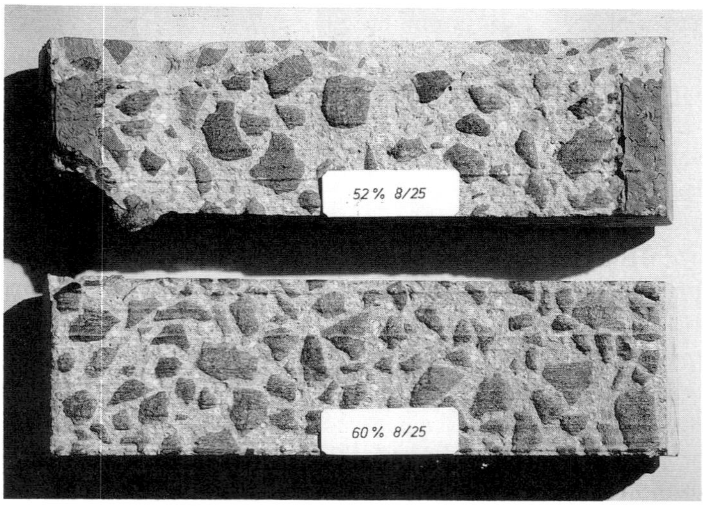

Fig. 3.27 (below left). Abrasive wear due to the scraping and percussive effects of studded tyres

Fig. 3.28 (below right). Wear due to the sliding action of an abrasive disc

of water, vapour-filled bubbles develop in this zone. If the bubbles stream to zones where the static pressure exceeds the vapour pressure of water, vapour condenses in the bubbles and the bubbles collapse suddenly. This implosion causes impact and pressure waves to develop, similar to those caused by explosions. This process is called cavitation, and results in damage similar to pitting and excavations.

Cavitation or similar impact and pressure waves occur when water hits limiting surfaces with a high velocity. Right-angled surfaces constitute an extreme case of this.

3.3.2. Influencing parameters

The abrasive wear resistance of concrete is borne by the coarse aggregates, which protect the less wear-resistant mortar against mechanical action, whether in air or in water. In contrast, wear resistance against cavitation is borne by the fine-grained mortar.

4. Chemical processes in concrete

4.1. Chemical attack on concrete

The durability of a concrete structure will often be determined by the rate at which the concrete is decomposed as a result of chemical reaction. With all these reactions, aggressive substances (ions and molecules) are being transported from somewhere, mainly from the environment, to — for this substance — a reactive substance in the concrete. However, even if the aggressive substance is already present in the concrete, it has to be transported in the direction of the reactive substance for the reaction to take place; if no transport takes place, there will be no reaction.

A precondition for chemical reactions to take place within the concrete at a rate which has any importance in practice is the presence of water in some form (liquid or gas).

In general, the reaction between the aggressive substance and the reactive substance takes place as soon as the substances meet. However, because of the low rate of transport of the aggressive substances within and into the concrete, these reactions often may take many years to show their detrimental effect.

The accessibility of the reactive substance in the concrete is therefore the rate-determining factor when an aggressive substance enters. The rate-increasing effect of increasing temperature is mainly due to the effect on the transport rate (higher temperatures result in higher mobility of ions and molecules). Depending on the type of reaction, the accessibility will be determined by the permeability of still sound concrete or by the passivating layer of the reaction products.

The chemical reactions that may lead to a decrease in quality are well established. The most important are

(*a*) the reaction of acids, ammonium salts, magnesium salts and soft water with the hardened cement
(*b*) the reaction of sulphates with the aluminates in the concrete
(*c*) the reaction of alkalis with reactive aggregates in the concrete.

A chemical reaction within the concrete increasing the risk of reinforcement corrosion is the reaction between calcium compounds, primarily $Ca(OH)_2$ and CO_2. This leads to carbonation of the concrete, causing a decrease in alkalinity. This mechanism is dealt with in section 6.2.2.

4.2. Acid attack

The action of acids (as the aggressive substance) on the hardened concrete (as the reactive substance) is the conversion of the calcium compounds (calcium hydroxide, calcium silicate hydrate and calcium aluminate hydrate) to the calcium salts of the attacking acid. The action of hydrochloric acid leads to the formation of calcium chloride, which is very soluble; sulphuric acid gives calcium sulphate, which precipitates as gypsum; and nitric acid gives calcium nitrate, which is very soluble. With organic acids, the result is the same: the action of lactic acid leads to calcium lactates; acetic acid gives calcium acetate, and so on.

As a result of the reactions, the structure of the hardened cement is destroyed (Fig. 4.1).

The rate of reaction of the different acids with concrete is determined not so much by the aggressiveness of the attacking acid, but more by the solubility of the resulting calcium salt. The less soluble the salt (if it is not carried away by other actions), the stronger will be its passivating effect. If the calcium salt is soluble, then the reaction rate will be determined largely by the rate at which the calcium salt is dissolved.

An important and generally valid condition governing deleterious chemical reactions is that the rate of deterioration caused by an aggressive chemical

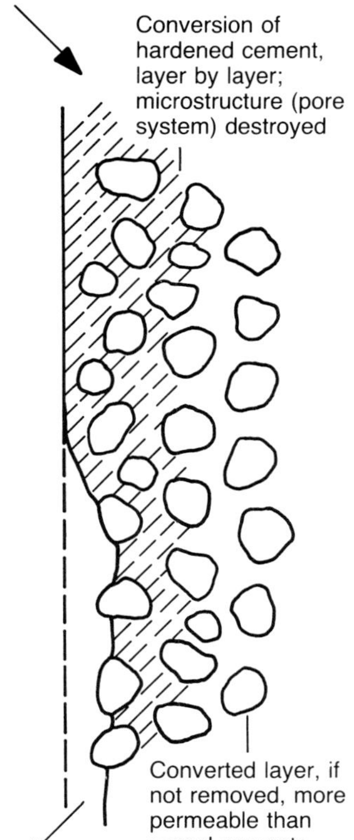

Fig. 4.1 (left). Effect of acid attack

Fig. 4.2 (below). Effect of sulphate attack

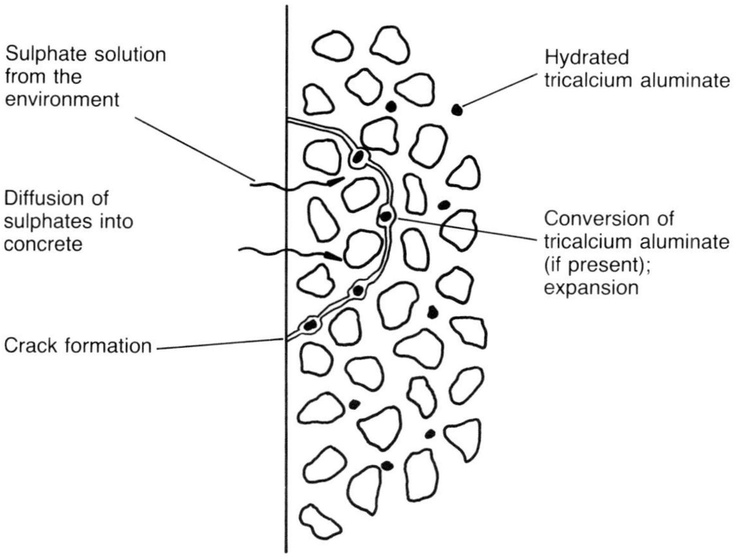

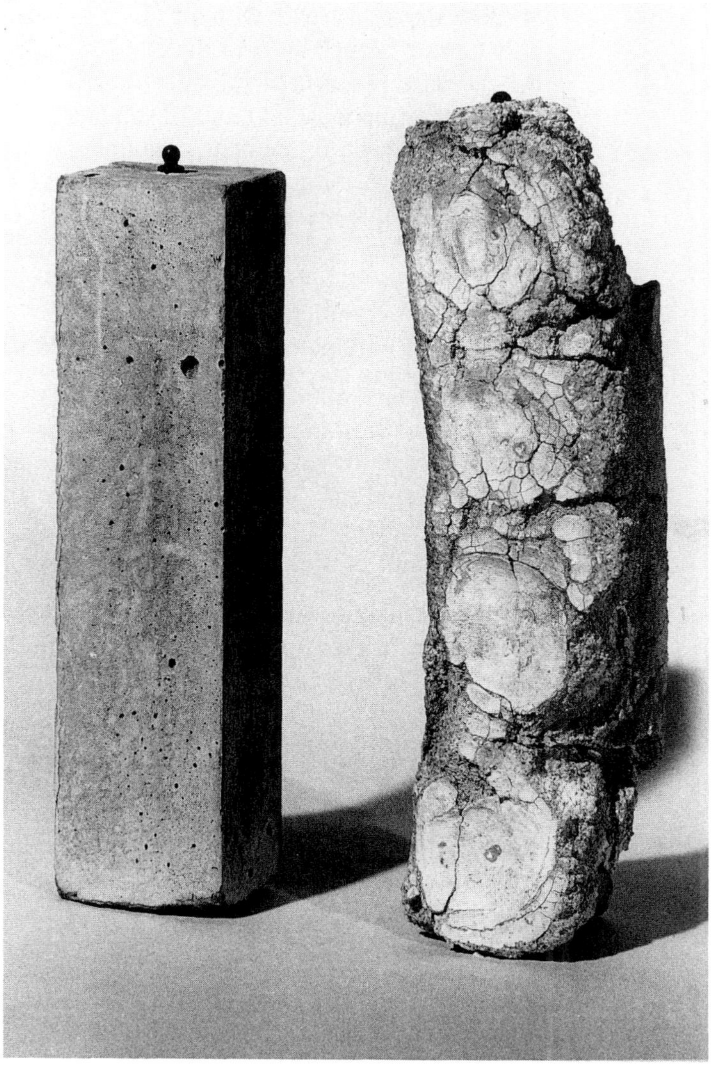

Fig. 4.3. Cracking due to sulphate attack

attack is much higher in a flowing solution than in a stagnant solution.

Magnesium and ammonium salts react in the same manner as the equivalent acids, so ammonium chloride will react as the free hydrochloric acid and ammonium nitrate as the free nitric acid. The only difference between the reaction of these two salts and the free acids is that in the former case magnesium hydroxide, and in the latter ammonium, is liberated.

Soft water merely dissolves the calcium compounds, as do the acids. The result is, again, the destruction of the hardened cement.

Regardless of the rate of reaction, the first thing that one should always calculate when discussing the possibility of acid attack or attack by magnesium salts, ammonium salts or soft water is the amount of substance the concrete comes into contact with. From this, one can calculate what the maximum loss of surface with time is, assuming a complete conversion of the acid into the calcium salt. It follows, for instance, that the amount of hardened cement that can be converted by acid rain is negligible, because the amount of acid falling each year is low compared with the buffering capacity of the concrete surface layer.

It should be realized that there is a fundamental difference between attack by acids and attack by sulphates and alkalis. In the former case, there is a complete conversion of the hardened cement, thus destroying the pore system. With acid attack, the permeability of the sound concrete is therefore of minor importance. With the other types of attack described below, the permeability of the sound concrete is of the utmost importance.

4.3. Sulphate attack

In contrast to acid attack, where the pore system as a whole is destroyed because the acids react with all the components in cement, sulphate attacks only certain components in the cement. Sulphate attack is characterized by the chemical reaction of sulphate ions (as the aggressive substance) with the aluminate component and ions of sulphate, calcium and hydroxyl of hardened Portland cement or cement containing Portland clinker (as the reactive substances), forming mainly ettringite and to a lesser extent gypsum.

The reaction between these substances, if enough water is present, causes expansion of the concrete, leading to cracking with an irregular pattern (Figs 4.2, 4.3). This gives easier access to further penetration, and so the process continues to complete disintegration.

The main parameters influencing the expansion in practice are

(*a*) exposure conditions, i.e. severity of attack (amount of aggressive substance)
(*b*) accessibility, i.e. permeability of concrete (rate of transport)
(*c*) susceptibility of concrete, i.e. type of cement (amount of reactive substance)
(*d*) amount of water available.

Concrete may to some extent be protected against sulphate attack, either by choosing a type of cement that is impervious to sulphate attack or by ensuring a sufficient degree of impermeability.

4.3.1. Exposure conditions

Exposure conditions may be modified by the presence of constituents other than sulphate and may have to be taken into consideration. An important example of this is the moderating influence of chloride ions caused by the preferential formation of chloro-aluminate (Fridell salt), which does not lead to detrimental expansion. Due to this mechanism sea water, which should be classified as highly aggressive according to its high sulphate content, is only moderately aggressive. Therefore, sea water, being of great importance as an exposure medium, is classified separately (see chapter 9 and section 13.4).

4.3.2. Accessibility of concrete

The degree of impermeability needed for a concrete to be sulphate resistant may be expressed as limiting values for depth of water penetration over a fixed period of time. For practical purposes, this is often translated into limiting values for W/C ratio or concrete quality. This holds true only for concrete with closed texture and does not account for shortcomings in the surface quality caused by local segregation and lack of curing.

Limiting values for water penetration and so on in highly aggressive media are still under discussion.

4.3.3. Cement type

The different types of cement may be classified according to their ability to resist sulphate attack. The American Society for Testing and Materials[8] limits aluminates to a maximum of 8% for moderate sulphate resistance (MSR) and to a maximum of 5% for high sulphate resistance (HSR). In Europe, a limit of 3% is generally accepted for (high) sulphate resistance.

Recent research has unanimously shown the good behaviour of blended cement. Several national standards recognize Portland blast-furnace cement with a minimum of 65% slag as HSR. The introduction of the MSR class allows due appreciation of other blended cements containing granulated slag or other pozzolanic material, either natural or synthetic (fly ash and silica fume).

It is important to realize that classification of cements for sulphate resistance only takes sulphate resistance as such into consideration. In cases of combined attack, other factors may influence the choice of cement. An example is the different behaviour of low alumina Portland cement and Portland blast-furnace cement with a high slag content. Both are HSR, but they have a very different permeability for chloride ions (as in sea water or due to de-icing salt); low alumina Portland cement results in the highest permeability towards chloride ions. This must be taken into consideration if corrosion of reinforcement is at stake.

4.4. Alkali attack

4.4.1. Alkali-silica reaction

The mechanism of alkali attack resembles that of sulphate attack more than acid attack, because the attack is only on certain substances in the concrete. The difference between sulphate attack and alkali attack is that the reactive

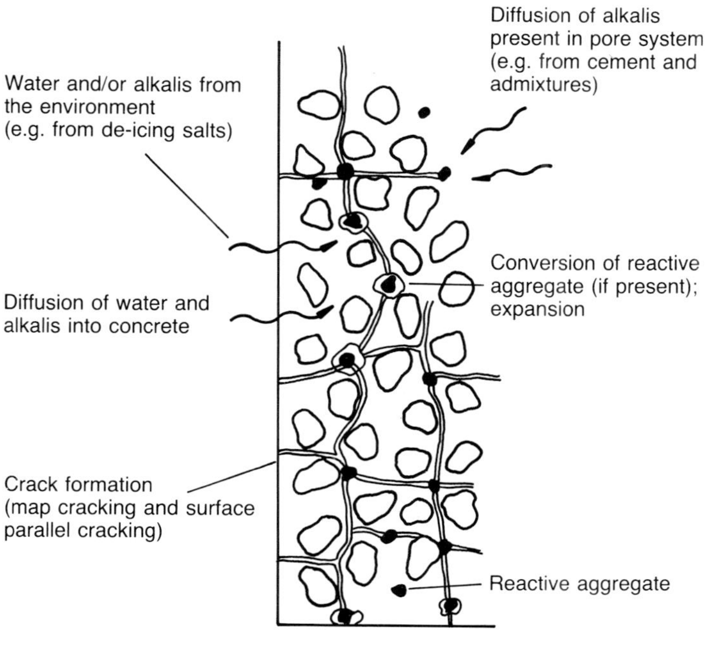

Fig. 4.4. Effect of alkali-silica reaction

Fig. 4.5 (left). Cracking due to alkali-silica reaction

Fig. 4.6 (right). Weeping of alkali-silica gel

substance in the former case is in the cement, and in the latter in the aggregates.

The alkaline solution in concrete pores is always lime-saturated and contains varying amounts of sodium and potassium ions. Silica-containing aggregates may be attacked by alkaline solutions. This may lead to destructive expansion (Fig. 4.4). Visible concrete damage starts with small surface cracks in an irregular pattern (map cracking), followed eventually by complete disintegration (Fig. 4.5). General expansion develops in the direction of least resistance, giving parallel surface crack patterns developing inward from the surface (for slabs), or cracking parallel to compression trajectories for compressed members (for columns or prestressed members). Other typical manifestations are pop-outs and weeping of glassy pearls of varying composition (Fig. 4.6).

So far, there has been no full explanation as to why the formation of alkali-silicate leads to expansion. The main parameters influencing the expansion in practice are

(*a*) the reactivity of the aggregate, which is based on the presence of amorphous or partly crystallized silica
(*b*) the amount and grain size of reactive aggregate
(*c*) alkali and calcium concentrations in the pore water (internal amount of aggressive substances)
(*d*) the type of cement (rate of transport)
(*e*) exposure conditions (external amount of aggressive substances)
(*f*) the amount of water available.

4.4.2. *Alkali-carbonate reaction*

Carbonate minerals may also be susceptible to alkaline attack. In dolomite or magnesium-containing limestone, the reaction may produce magnesium hydroxide. This 'dedolomitization' may lead to map cracking, resulting ultimately in the complete destruction of the concrete.

As far as is known, this type of reaction has not occurred in Europe.

4.4.3. *Susceptibility of aggregate*

4.4.3.1. Alkali-silica reaction. The presence of reactive silica is one limiting factor. Assessment of reactivity is difficult, however, and a method that gives satisfying results for all potential aggregates under all possible circumstances is not yet available.

Deleteriousness of alkali reaction does not simply increase with the amount of reactive aggregate; at a certain fraction, the expansion reaches a maximum. Generally, this fraction amounts to no more than a few percent; it is also influenced by cement type and concrete mix. Furthermore, the deleteriousness is dependent on the grain size of the reactive material.

Instead of absolute levels of expansion, it may be better to consider the rate of expansion. Observation until expansion becomes negligible makes it possible to adjust observation time for an individual type of aggregate.

4.4.3.2. Alkali-carbonate reaction. Assessment of alkali-carbonate reactivity, which is far less common than alkali-silicate reactivity, generally follows the same lines. Petrographic distinction of potentially dangerous material is easily made. A deleterious degree of expansion is only reached in the presence of clayey components, possibly expressed as alumina content.

4.4.4. Alkali content

As alkali concentration in pore water is a decisive factor, the alkali content of concrete at any given time is important. Free alkali is mainly supplied by the cement. Other sources, especially the influx of alkali-containing water into hardened concrete, may have to be taken into consideration.

4.4.5. Cement type

Portland cements with limited alkali content are special cements with respect to alkali-aggregate reactivity, and have been used as such for many years.

The use of blended cements normally causes a decrease in both the alkali and the calcium concentration together with a decreased permeability.

Certain standards allow rather high limits for alkali content for blast-furnace slag cements (with limits depending on slag content).

4.4.6. Exposure conditions

Although largely neglected in the existing standards and recommendations, exposure conditions certainly play a role and may be responsible for the great difference in rate of deterioration of concrete with the same amount and type of reactive aggregate.

For concrete design, judgement of aggregates is based on test results at constant and high humidity. It is known that intermittent drying and wetting may lead to greater expansion.

A practical implication of the influence of exposure is the possibility of retarding or even preventing a progressing deterioration by waterproofing the concrete.

5. Biological processes in concrete

Growth on concrete structures may lead to mechanical deterioration caused by lichen, moss, algae, and roots of plants and trees penetrating into the concrete at cracks and weak spots, resulting in bursting forces causing increased cracking and deterioration. Such growth may also retain water on the concrete surface, leading to a high moisture content of the concrete with subsequent increased risk of deterioration due to freezing. Furthermore, microgrowth may cause chemical attacks by developing humic acid, which will dissolve the cement paste.

In practice, the most important type of biological attack on concrete occurs in sewer systems. In anaerobic (oxygen-free) conditions, hydrogen sulphide (which is itself not very aggressive for concrete) can be formed from sulphate or from proteins in the sewage. After escape of this hydrogen sulphide from the solution (depending on chemical equilibrium and turbulence), it may be oxidized by bacteriological action to form sulphuric acid, thus resulting in an acid and sulphate attack on the concrete above the water level (Fig. 5.1).

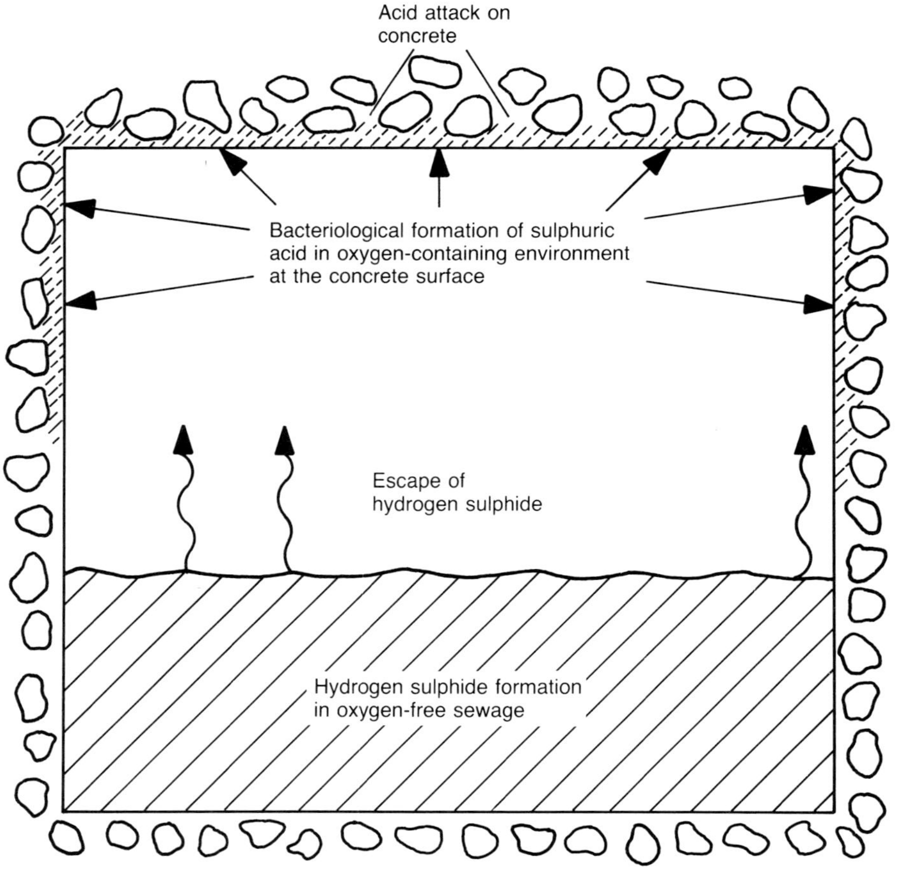

Fig. 5.1. Biological attack in sewer systems

A detailed description of the process and measures to be taken has been given by Thistlethwayte.[9]

In concrete structures in the sea, marine growth may help to protect the structure. The plants consume oxygen before it can diffuse into the concrete, thus preventing it from taking part in a corrosion process on the reinforcement.

6. Reinforcement

6.1. Protection of steel in concrete: normal situation

Steel in concrete is protected against corrosion by passivation. This passivation is due to the alkalinity of concrete: the pH of the pore water runs up to greater than 12·5. With such high pH-values, a microscopic oxide layer is formed on the steel surface — the 'passive' film — which impedes the dissolution of iron. Corrosion of reinforcement is thus impossible, even if all other preconditions for corrosion (mainly the presence of moisture and oxygen) are fulfilled.

6.2. Mechanisms of corrosion and corrosion protection

6.2.1. Processes and effects

Due to carbonation of the concrete or by the action of chloride ions, the passive film may be destroyed locally or over greater surface areas. A third mechanism is a reduction of alkalinity due to the leaching out of alkalis by streaming water. In practice, this may happen in the region of weak points of the structure (e.g. leaky construction joints and wide cracks) in combination with bad concrete quality (gravel pockets, high W/C ratio). If the pH of concrete drops below 9 at the reinforcement, or if the chloride content exceeds a critical value, the passive film and the corrosion protection will be lost.

Consequently, corrosion of reinforcement is possible, if sufficient moisture and oxygen are available. This can be assumed to be the case for structures in the open air.

The principles of passivation and depassivation hold true for reinforcing as well as for prestressing steels.

6.2.2. Carbonation of concrete

Concrete is a porous material, and the CO_2 in the air may therefore penetrate via the pores to the interior of the concrete. There, a chemical reaction will take place with the calcium hydroxide. In very simplified terms, the chemical reaction may be described as

$$Ca(OH)_2 + CO_2 \rightarrow CaCO_3 + H_2O$$

As it is mainly the $Ca(OH)_2$ that causes the high pH of the concrete to develop, the pH will drop below 9 after the concrete has been totally carbonated.

As already mentioned, the CO_2 penetrates from the surface to the interior of the concrete. Consequently, the carbonation starts from the concrete surface and penetrates slowly to the interior of the concrete. The rate-determining process is the diffusion of CO_2 into concrete. Roughly simplified, therefore, the rate of carbonation (increase of carbonation depth with time), follows a square-root time law (Fig. 6.1).

The concrete quality parameter in relation to carbonation is the permeability, which for a given environment depends on the pore structure. Diffusion of CO_2 is only possible in air-filled pores. For this reason, totally water-saturated concrete will not carbonate.

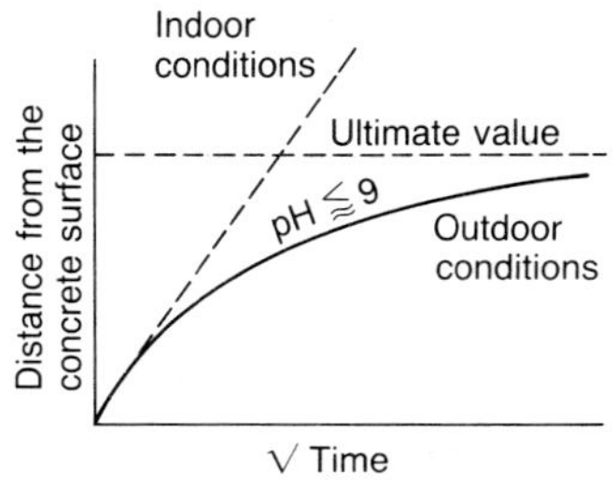

Fig. 6.1. Rate of carbonation (increase of carbonation depth with time); the ultimate value decreases with the permeability of the concrete, the amount of carbonizable substance and increasing environmental humidity

6.2.3. Penetration of chlorides into concrete

Besides CO_2, chloride ions (originating from sea water or de-icing salt) may penetrate through the pores to the interior of the concrete. Chloride intrusion is due to either diffusion taking place in totally or partially water-filled pores or capillary suction of chloride-containing water.

Cement has a certain chemical and physical binding capacity for chloride ions (forming Fridell salt), depending on the chloride concentration in the pore water. However, not all the chlorides can be bound. There will always

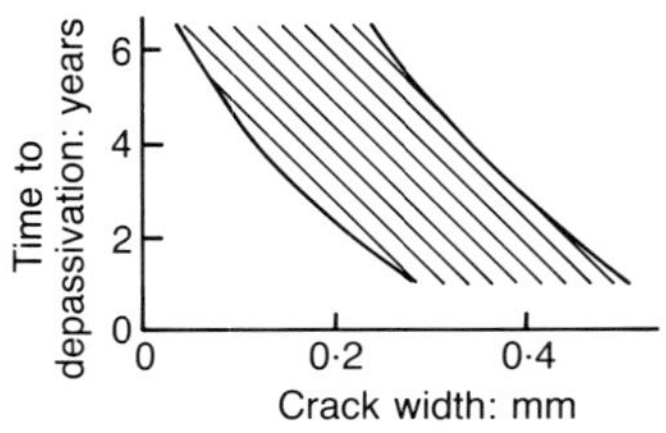

Fig. 6.2. Relationship between depassivation time and crack width; the scatter depends on the environment, the cover and the nature of any deposits

exist a dissolution equilibrium between bound chlorides and free chloride ions in the pore water. Only the free chloride ions are relevant to the corrosion of the reinforcement. It is important to note, therefore, that after carbonation of concrete bound chlorides are released again, so that the chloride content in the pore water, and consequently the risk of corrosion due to chlorides, will increase considerably. The critical chloride concentration at which corrosion will occur depends on many parameters (see section 12.2.1.7).

As a result of the diffusion process, the chloride concentration will decrease from the surface to the interior of the concrete. To a rough approximation, the penetration depth again follows a square-root time law. However, exact calculations and observations in practice show that the penetration rate is slower than that which results from the square-root time law. The main reason for this effect is the change of pore size distribution with time due to the continuing hydration process.

Due to wetting and drying of the concrete surface with chloride-containing water, an enrichment of chlorides in the surface layer is possible. At the beginning of the wetting period, a relatively large amount of chloride-containing water will penetrate into the concrete by capillary suction. During the drying period, the water dries out and the chlorides remain in the concrete. This process may cause a high enrichment of chlorides in the drying and wetting zone of a concrete. Therefore, the water penetration depth of a concrete and the permeability of the surface layer, respectively, are of great importance, especially in relation to the thickness of the concrete cover.

6.2.4. Depassivation in the area of cracks crossing the reinforcement

Both CO_2 and chlorides may penetrate to the steel surface through cracks some order of magnitudes faster than through uncracked concrete. The time taken for depassivation depends on the crack widths; however, the times involved are negligible compared with the lifetime of reinforced concrete structures (Fig. 6.2).

In the case of post-tensioned structures, durable passivation of the prestressing steel can be assumed if

(*a*) cover to the ducts exceeds 5 cm
(*b*) crack widths are less than 0·2 mm at the concrete surface
(*c*) ducts are thoroughly and completely grouted
(*d*) chlorides are absent.

6.2.5. Corrosion of reinforcement

As a simplified model, the corrosion process can be separated into two single processes: the cathodic and the anodic process (Fig. 6.3).

Fig. 6.3. Simplified model for corrosion of reinforcement in concrete

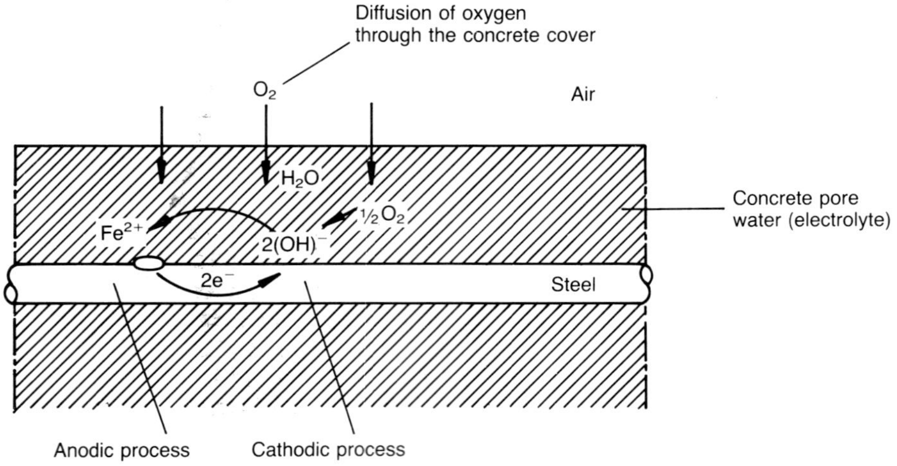

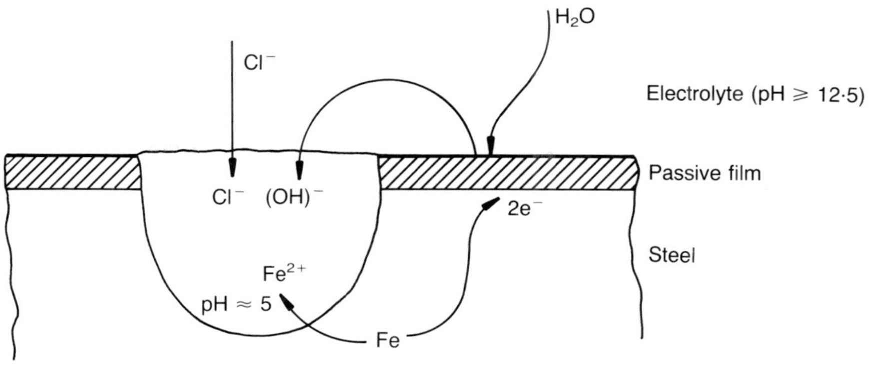

Fig. 6.4. Pitting corrosion caused by chlorides

The anodic process is the dissolution of iron. Positively charged iron ions pass into solution

$$Fe \rightarrow Fe^{2+} + 2e^-$$

The surplus electrons in the steel will combine at the cathode with water and oxygen to form hydroxyl ions

$$2e^- + \tfrac{1}{2}O_2 + H_2O \rightarrow 2(OH)^-$$

After some intermediate stages, the iron and hydroxide ions will combine to form rust which, at least theoretically, can be written as Fe_2O_3 (under practical conditions, rust products are more or less water-containing compounds). This means that only oxygen is consumed to form rust products. This oxygen must normally diffuse through the concrete cover towards the reinforcement. Water is only necessary to enable the electrolytic process to take place.

As a consequence of the interrelations described, corrosion will not occur either in dry concrete (where the electrolytic process is impeded) or in water-saturated concrete (where oxygen cannot penetrate), even if the passive layer at the surface of the reinforcement has been destroyed. The highest corrosion rate will occur in concrete surface layers subjected to highly changing wetting and drying conditions.

In the anodic areas, the passive film must be destroyed; the cathodic process, however, can take place even if the passive layer is intact. In the case of chloride corrosion, this effect causes pitting corrosion, because the passive layer will be dissolved only over small surface areas, so that small anodic areas and huge cathodic areas will exist on the surface — a fact that causes substantial local reductions in sections of the reinforcement. In addition, the chloride ions will act as a catalyst in the pit and accelerate the dissolution of iron in the anodically-acting pit (Fig. 6.4).

At the steel surface, anodically and cathodically acting areas may be situated either close together (microcell corrosion) or at locally separated places

Fig. 6.5. Example of macrocell corrosion

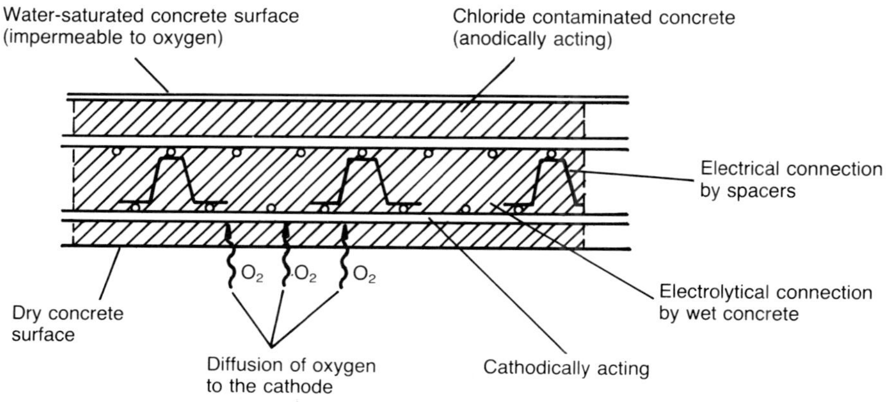

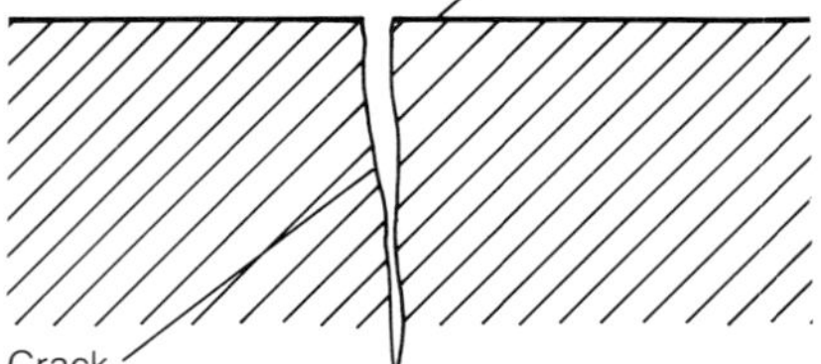

Fig. 6.6. *Stress corrosion cracking*

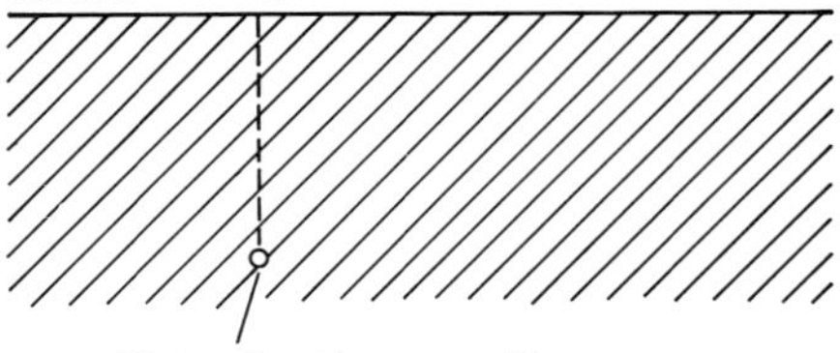

Fig. 6.7 *(far right). Hydrogen embrittlement*

(macrocell corrosion) even over relatively great distances. Consequently, corrosion may occur in areas of the structure where the direct access of oxygen to the surface of the reinforcement is impeded, if the concrete is wet enough to render the electrolytical connection possible (Fig. 6.5).

6.2.6. *Stress corrosion cracking and hydrogen embrittlement*

In addition to the corrosion processes described in the previous section, failures of a brittle nature, caused by corrosion, may occur in prestressing steel.

Very localized anodic processes may lead to cracking due to high permanent stresses, if the steel is sensitive to this type of failure. During the crack propagation stage, the anodic process takes place at the root of the crack (Fig. 6.6). This type of brittle cracking is called stress corrosion cracking (SCC).

The second type of brittle failure is the consequence of a cathodic process. Under certain conditions, atomic hydrogen is developed during the cathodic process as an intermediate product and may penetrate into the steel. The recombination to molecular hydrogen within the steel leads to high local internal pressure and may, consequently, lead to cracking (Fig. 6.7). This type of failure is called hydrogen embrittlement (HE).

Both types of failure are consequences of at least local depassivation, and will not occur if the prestressing steel is totally surrounded by sound hardened concrete or cement grout.

As a special case, atomic hydrogen may develop at zinc-coated metal surfaces in fresh concrete or grout. Therefore, the use of galvanized (zinc-coated) ducts leads to a high risk of hydrogen embrittlement provided that the prestressing steel is in electrical contact with the duct. However, for two reasons this risk is only temporary: the evolution of hydrogen will come to a stop when the concrete or the grout has hardened well; and if hydrogen penetrates into the steel without causing a failure, it will diffuse out again, thus relieving the local bursting pressure and reducing the risk.

6.2.7. *Influence of cracks*

In the region of cracks, carbonation and chlorides tend to penetrate faster towards the reinforcement than in uncracked concrete. In the case of normal crack widths at the concrete surface of up to 0·4 mm, self-healing as a result of calcium, dirt and rust deposits within the cracks can frequently be observed. In this case, any on-going corrosion at the reinforcement is likely to come to a halt.

The thickness of the concrete cover is of major importance with regard to the influence of cracks. The crack widths (if they are less than 0·4 mm) are less important.

6.2.8. *Corrosion processes in the region of cracks*

If carbonation or chlorides have reached the reinforcement, depassivation of the reinforcement may occur (see section 6.2.1). Corrosion current measurements show that normally macrocell corrosion occurs, the steel in the crack region acting anodically while the cathodic process takes place in the uncracked areas beside the cracks (Fig. 6.8). In this process the crack widths are of minor importance after depassivation, because the cathodic process is the main rate-determining factor.

Results of exposure tests and site inspections confirm these theoretical findings. The influence of crack width on the corrosion rate at the reinforcement turns out to be relatively small within the common range of crack widths (up to 0·4 mm). Of substantially greater importance is the thickness of the concrete cover.

Cracks oriented transverse to the reinforcement are less harmful than longitudinal cracks. This is due to the fact that in the case of transverse cracks, corrosion is confined to a small surface area, so that there is no risk of spalling of the concrete cover.

Cracks crossing the reinforcement may be harmful if horizontal concrete surfaces are directly affected by chloride-containing water. In such cases special protective measures, (e.g. sealing or lining of the concrete or coating of the reinforcement) should be provided. Limitation of crack widths cannot reduce the corrosion risk under these circumstances.

6.2.9. Effect of corrosion

The corrosion process may result in a reduction of cross-section of the reinforcement and splitting of the concrete cover. If the cross-section is reduced the load-bearing capacity of the steel decreases in a roughly linear fashion, whereas the elongation properties and the fatigue strength may be reduced more substantially by a small reduction in cross-section. This means that the latter two properties are much more sensitive to corrosion than the load-bearing capacity.

Rust has a substantially higher volume than steel — theoretically up to more than six times greater, depending on oxygen availability. This leads to splitting forces that may cause cracking and spalling. This effect of corrosion of reinforcement may lead to sudden failure, if longitudinal cracking along the bars occurs in the region of the bar anchorages.

When corrosion develops in environments with low availability of oxygen, the volume of the rust products may only be 50–200% greater than the volume of the steel. Such corrosion processes proceed slowly, and in special cases the rust products may diffuse into the voids and pores of the porous concrete

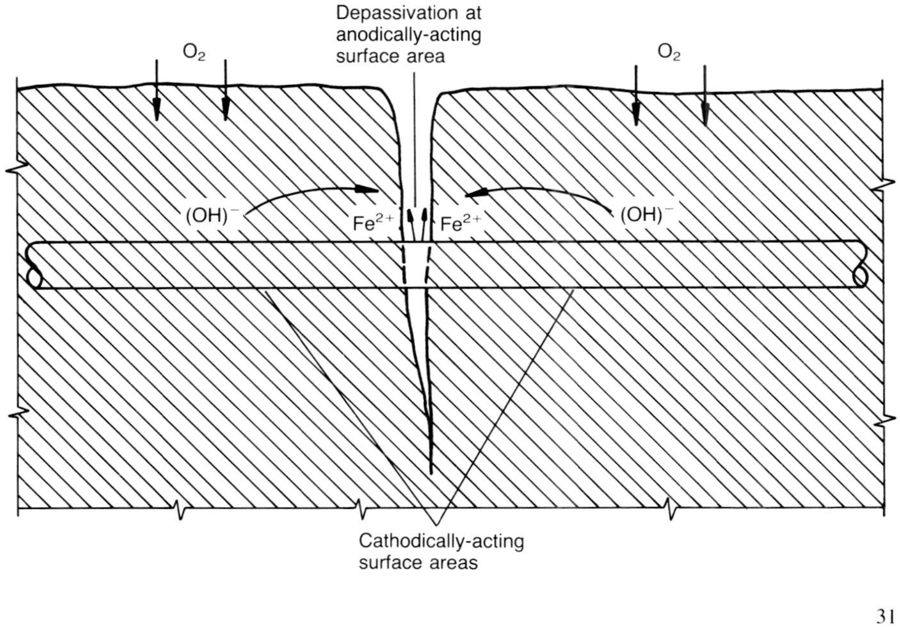

Fig. 6.8. Corrosion of reinforcement in the region of cracks

without causing cracking and spalling. In such rare cases serious corrosion may develop on the reinforcement without any visible warning, and a sudden failure may occur.

6.3. Influencing parameters

All the processes influencing corrosion of reinforcement are more or less controlled by transport processes

(*a*) carbonation: diffusion of CO_2 in air-filled pores
(*b*) penetration of chlorides: diffusion of chlorides in water-filled pores and capillary suction of chloride-containing water into air-filled pores
(*c*) corrosion of reinforcement: diffusion of oxygen in air-filled pores.

Therefore, the major parameter in connection with corrosion and protection of the reinforcement in both uncracked and cracked concrete is the quality of the concrete cover. This quality is defined in terms of the thickness and permeability of the concrete cover.

Another important parameter is the microclimate at the concrete surface (see section 6.3.5).

6.3.1. Thickness of concrete cover

As shown in section 6.2, carbonation and chlorides penetrate to the interior of concrete at a lower rate than would be given by a square-root time function. This means that if the concrete cover is halved, the critical state for incipient danger of corrosion will be reached in less than a quarter of the time (Fig. 6.9).

6.3.2. Permeability of concrete cover

6.3.2.1. Influence of W/C ratio. The water/cement ratio of concrete influences the permeability of concrete decisively. Particularly in cases where the W/C ratio exceeds 0·6, the permeability will increase considerably with W/C ratio, due to the increase in the capillary porosity. Figure 6.10 shows how the water permeability depends on the W/C ratio and the degree of hydration. In principle, the same basic influence of W/C ratio holds true for gas and ion permeability.

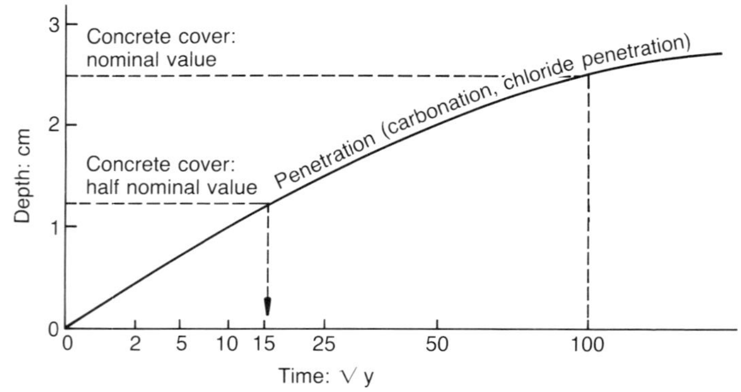

Fig. 6.9. Example of the effect of the thickness of the concrete cover. For the nominal concrete cover, carbonation reaches the surface of the reinforcement after 100 years. If the cover is reduced to half of the nominal thickness, the penetration occurs in only 15 years

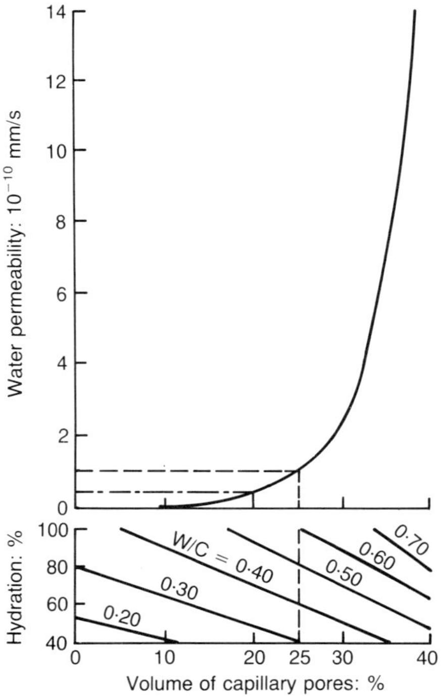

Fig. 6.10 (right). Influence of W/C ratio on permeability[10]

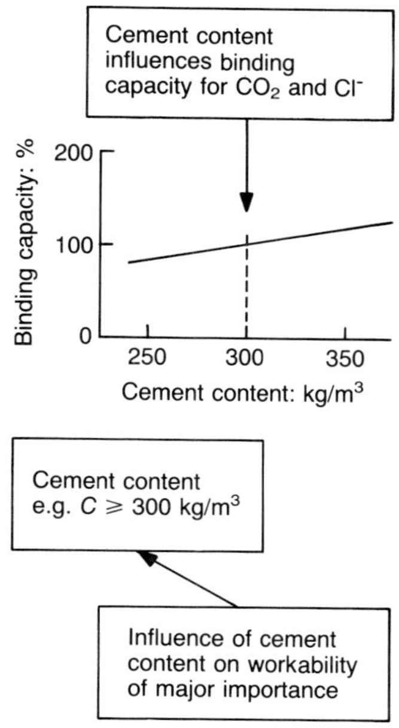

Fig. 6.11. Influence of the cement content on binding capacity

6.3.2.2. Influence of curing. If the concrete is insufficiently cured (i.e. the concrete surface dries early), the permeability of the surface layer of concrete may be increased by five to tenfold. The depth of the influenced layer depends on the grade of drying; however, it is often equal to or thicker than the concrete cover. Wind and high temperatures are very dangerous as far as early drying out of the concrete surface is concerned.

Curing measures taken after the first drying out of concrete are useless, because the hardening will hardly continue after having been interrupted once. Therefore, curing measures must begin immediately after concreting and are not to be interrupted.

The curing sensitivity increases with increasing W/C ratio and decreasing cement content. The influence of type of cement on curing sensitivity is discussed in section 6.3.4.

6.3.2.2. Influence of compaction. Poor compaction or gravel pockets tend to increase the permeability of concrete to such an extent that protection of the reinforcement no longer exists.

6.3.3. Cement content

With increasing cement content, the binding capacity of concrete both for CO_2 and Cl^- will be increased (Fig. 6.11). However, over the normal range of cement contents the penetration rates of carbonation and chlorides are influenced to a considerably lower extent by the cement content than by the W/C ratio, the quality of compaction, and curing. Nevertheless, the amount of cement is important in connection with the workability and, to a certain extent, with the curing sensitivity.

Normally, a cement content in the range of 300 kg/m^3 is sufficient to achieve a sufficiently low permeability and sufficient durability if the W/C ratio is kept below 0·5–0·6, depending on the environmental conditions (the presence or absence of chlorides) and provision of adequate curing. In cases where special care is taken to achieve a good quality concrete, a lower cement content may be sufficient.

An alternative means of ensuring sufficient concrete quality may be by specifying relatively high minimum strengths, differentiated according to the exposure classes.

6.3.4. Cement type

Generally, the most common composite and blended cements with natural pozzolanas, blast-furnace slag or fly ash have in common the properties of

(*a*) slow hardening at an early age
(*b*) distinct hardening later on.

This means that composite and blended cements are more curing-sensitive than Portland cements.

If the later hardening is ensured by adequate curing, a lower permeability of concrete can be achieved by using composite or blended cements rather than Portland cements. In this way, especially, the resistance against chloride penetration can be improved.

Whatever the type of cement, inadequate curing may lead to a poor quality (in terms of permeability and binding capacity) of the concrete cover. The sensitivity to curing is especially pronounced if cements with high percentages of blending agents (e.g. in excess of 50% slag, 15% fly ash or 8% silica fume) are used (Fig. 6.12).

In addition, the freezing-thawing resistance must be considered if highly blended cements are used.

6.3.5. Environment

In permanently dry environments (relative humidity less than 60%) the

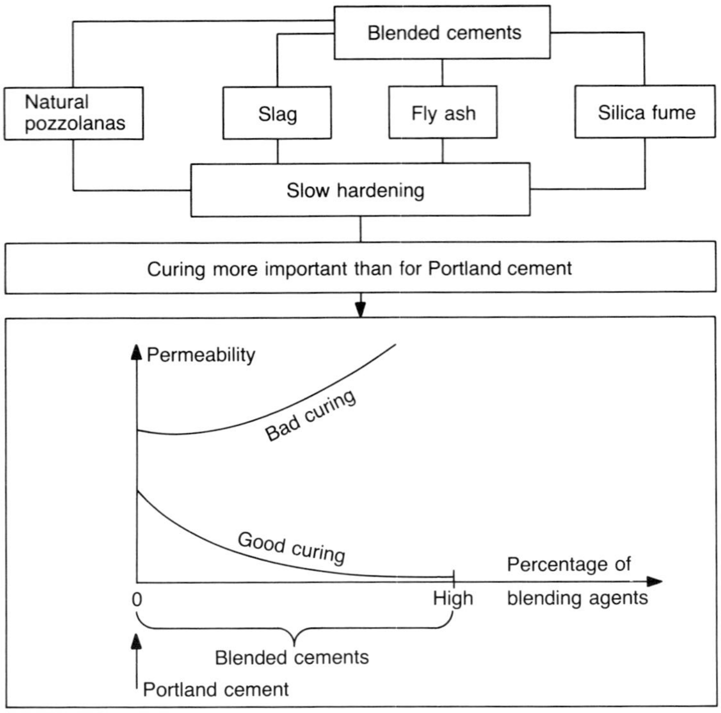

Fig. 6.12. Influence of the type of cement on permeability

corrosion risk is low, even if the concrete is carbonated, because the electrolytic process is impeded. In the case of high chloride content, corrosion may be possible even in dry environments due to hygroscopic effects, which increase the water content of the concrete.

In permanently water-saturated concrete the corrosion risk is low due to the lack of oxygen, even if the concrete is highly chloride-contaminated. However, the risk of separated anodically- and cathodically-acting steel surface areas must be taken into account if the structure or structural element is only partly saturated or immersed (see section 6.2.5).

The most favourable conditions for corrosion of steel in concrete are alternating wetting and drying combined with high temperatures. All processes involved are considerably accelerated with increasing temperature.

6.3.6. Conclusions

Of major importance for the quality of the outer concrete layer (i.e. the cover) are

(*a*) W/C ratio
(*b*) compaction
(*c*) curing.

The cement content mainly influences the workability, and thus, indirectly, the permeability and the curing sensitivity of the concrete cover.

The surface layer of concrete is especially susceptible to increased permeability caused by inadequate design and execution. In this case, any locally reduced concrete cover may reduce the durability of the structure considerably.

7. Environmental aggressivity

For deleterious processes to develop — for concrete as well as for reinforcement (reinforcing and/or prestressing) — interactions have to take place between the material in the structure and the environment. These interactions depend in type, intensity and timing on the material properties, especially the permeability (see section 6.3.2), the selected structural form, and the position of the reinforcement, and on the type and aggressiveness of the environment. The properties of the environments surrounding buildings should therefore be clarified with respect to their influence on durability.

The general atmospheric climate (or macroclimate) around buildings may be determined easily through traditional means, but has only minor importance for durability. Of decisive influence is the local climate within metres of the structure, or even the microclimate (millimetres or centimetres away), and the conditions around buried (e.g. foundations or piles) or submerged parts of the structure.

Unfortunately, no generally accepted method yet exists for rigorously defining environments with respect to their aggressivity towards concrete structures, i.e. towards the concrete and towards the reinforcement, whether prestressed or non-prestressed. There are many different categorizations of environments currently in use. In section 15.1.4.1 of the CEB–FIP model code[6] the following conditions of exposure are given

(*a*) mild
- (i) the interiors of buildings for normal habitation or for offices
- (ii) conditions where a high level of relative humidity is reached for only a short period in any one year (for example, relative humidity only exceeds 60% for less than 3 months in a year)

(*b*) moderate
- (i) the interiors of buildings where the humidity is high or where there is a risk of the temporary presence of corrosive vapour
- (ii) running water
- (iii) inclement weather in rural or urban atmospheric conditions without heavy condensation of aggressive gases
- (iv) ordinary soils

(*c*) severe
- (i) liquids containing slight amounts of acids, saline or strongly oxygenated waters
- (ii) corrosive gases or particularly corrosive soils
- (iii) corrosive industrial or maritime atmospheric conditions.

It is assumed that these categories correspond to slightly aggressive, moderately aggressive and highly aggressive environments in section 5.1 of the model code.[6]

This gives some guidance when estimating the durability risks associated with a given structure in a given environment.

In a recent draft CEN document[11] (prEN 206) a more comprehensive classification of environmental exposures has been presented. A proposal for an operational classification scheme is presented in chapter 9 of this design guide. There, the CEN proposal has been supplemented with a separate classification of environmental conditions aggressive to the reinforcement.

Clearly, it is not easy to decide on the conditions of exposure of particular elements in particular environments. However, the factors described in the following three sections are known to have a dominating influence on the aggressivity of a particular environment.

Table 7.1. Influence of moisture state on durability processes

Effective relative humidity	Process*				
	Carbonation	Corrosion of steel		Frost attack	Chemical attack
		In carbonated concrete	In chloride contaminated concrete		
Very low (<45%)	1	0	0	0	0
Low (45–65%)	3	1	1	0	0
Medium (65–85%)	2	3	3	0	0
High (85–98%)	1	2	3	2	1
Saturated (>98%)	0	1	1	3	3

* 0 = insignificant risk; 1 = slight risk; 2 = medium risk; 3 = high risk.

7.1. Availability of moisture

All deterioration processes require water: the important factor is the moisture state in the concrete rather than that of the surrounding atmosphere. Under steady conditions these will be constant but under varying conditions concrete takes water in from the environment more rapidly than it loses it (see section 2.5), and so the internal average humidity tends to be higher than the average ambient humidity.

This principle also holds true where members are subject to wetting and drying: frequent wetting, as in tidal regions, can maintain concrete in a saturated condition.

Table 7.1 indicates the influence of effective humidity on various processes related to durability.

As an example of how much influence the presence of moisture has on corrosion of reinforcement, the solid line in Fig. 7.1 indicates in gross terms the relative risk of corrosion damage dependent on the mean annual effective relative humidity (i.e. the humidity in the pores of the concrete) in a normal environment. The scale of aggressivity has been defined so that aggressivity is directly proportional to the cover required to produce a uniform risk of damage (i.e. twice the aggressivity will require twice the cover).

7.2. Presence of aggressive substances in moisture

Common examples of aggressive substances which may be present in moisture are

(*a*) carbon dioxide — necessary for carbonation
(*b*) oxygen — necessary for corrosion
(*c*) chlorides — promote corrosion

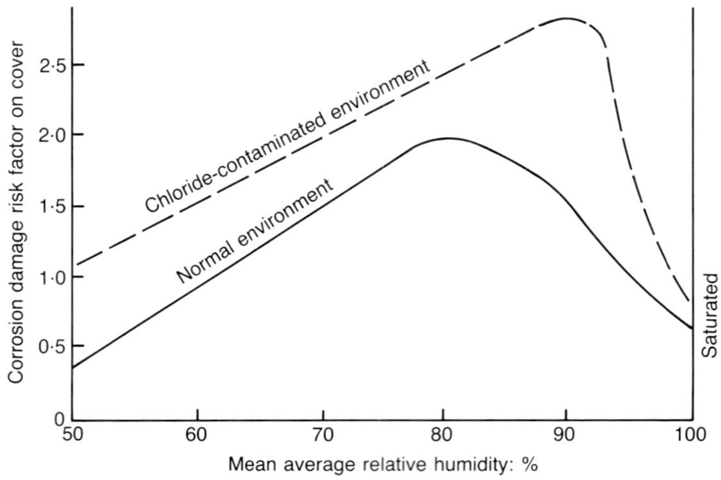

Fig. 7.1. Influence of moisture on corrosion risk relative to cover

(*d*) acids — dissolve cement
(*e*) sulphates — give expansive reaction with cement
(*f*) alkalis — give expansive reaction with aggregate.

As an example, Fig. 7.1 indicates in gross terms the increased risk of corrosion damage when the environment is chloride-contaminated compared with the risk in normal environments. It should be emphasized that the abscissa represents the effective relative humidity, i.e. the relative humidity within the concrete. For a given atmospheric environment the water content in concrete will be higher when chlorides are present, due to their hygroscopic effect. This accounts, for example, for the heavy corrosion encountered with chloride-contaminated concrete placed indoors in permanently air-conditioned rooms where temperatures are average (20°C) and relative humidity is low (50–60%), such as in the Middle East.

7.3. Temperature level

The influence of temperature tends to be ignored in definitions of aggressivity, but is very important, as chemical reactions are accelerated by increases in temperature. A simple rule-of-thumb is that an increase in temperature of 10°C causes a doubling of the rate of reaction. This factor alone makes tropical environments considerably more aggressive than, for example, Northern European climates. Figure 7.2 shows the influence of temperature on environmental aggressivity in cases where the thickness of concrete cover is the rate-determining factor. The scale is defined such that the aggressivity is directly proportional to the cover required to produce a uniform risk of damage.

The availability of moisture, the presence of aggressive substances in moisture, and the temperature level are the main considerations in characterizing a particular environment. In doing this, however, it is necessary to consider the interaction between some of the effects. One example will be considered here, as it is of considerable importance: the corrosion of reinforcement where the passivity of the steel has been destroyed by carbonation and not by chlorides.

Carbonation is most rapid when the relative humidity is in the region of 50–60%. Below this there is insufficient moisture for the reaction to be significant, and above this the water in the pores increasingly inhibits the ingress of carbon dioxide until, at about 95%, carbonation is almost completely inhibited. The rate of corrosion, however, is very low when the relative humidity is in the 50–60% region and highest when the humidity is 90–95%.

Fig. 7.2. Influence of temperature on environmental aggressivity relative to cover

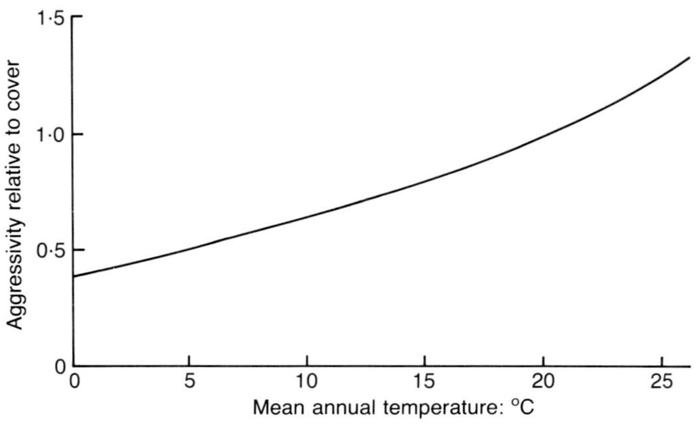

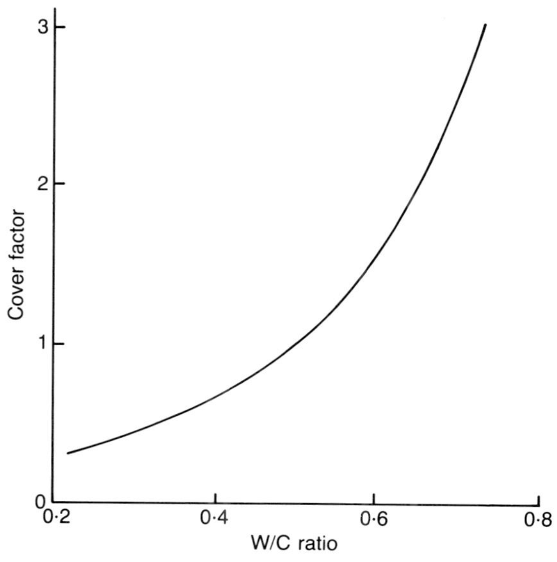

Fig. 7.3 (right). Influence of W/C ratio on permeability relative to the efficiency of cover in protecting reinforcement

Table 7.2. Minimum concrete cover in mm for reinforcement of low susceptibility to corrosion (from table 5.2 of ref. 6)

Conditions of exposure	Grade of concrete					
	C12, C16, C20		C25, C30, C35		C40, C45, C50	
	General case	Slabs shells	General case	Slabs shells	General case	Slabs shells
Mild	20	15	15	15	15	15
Moderate	30	25	25	20	20	15
Severe	40	35	35	30	30	25

Above this, the corrosion rate drops rapidly to a very low value for saturated concrete, due to lack of oxygen.

For a durability failure to occur, carbonation must have reached the steel over a substantial area and an unacceptable amount of corrosion must have occurred. It follows that the risk of corrosion damage will be low at the humidity levels corresponding to the maximum carbonation; maximum corrosion rates and the highest risk of corrosion damage will correspond to some intermediate humidity.

The mechanisms by which chlorides commonly penetrate to the reinforcement are quite different from carbonation, and the effect of humidity is irrelevant. The risk of corrosion damage in the presence of chlorides will therefore be expected to be directly related to humidity in the same way as is corrosion rate.

7.4. Concrete cover

The susceptibility of reinforcement to corrosion, together with the thickness of the concrete cover protecting the reinforcement and the quality (i.e. the permeability and alkalinity) of the cover, interact with the environment in a way which determines whether the environment is aggressive to the reinforcement or not. Section 5.1 of the CEB-FIP model code[6] gives the covers in Table 7.2 for reinforced concrete in various environments.

In chapter 9 of this design guide an enlargement of Table 7.2 is proposed to cope with the enlarged and more comprehensive classification of exposure conditions for the reinforcement.

The ability of the concrete in the cover to protect the reinforcement depends to a large extent on its low permeability to aggressive substances in liquid or gaseous form. The permeability is directly related to the W/C ratio, and depends furthermore on correct execution and curing. Figure 7.3 gives a gross indication of how much the cover should be increased with increased W/C ratio in order to maintain the same low risk of corrosion, i.e. maintain approximately the same service life. However, trading of cover against W/C ratio or curing should be limited to avoid error-sensitive solutions.

8. Scope of the recommendations

As an introduction to the recommendations, it may help to clarify the objectives of design for durability.

Concrete structures are designed and constructed with the aim of satisfying a set of functional requirements over a certain period of time without causing unexpected costs for maintenance and repair. This period of time constitutes the anticipated lifetime or design service life of the structure. Such a concept is implicit in all design rules, including the model code,[6] but an actual figure for this design life is rarely stated explicitly. Exceptions to this are the British bridge code,[12] which specifies a design life of 120 years, and the British code for farm buildings,[13] which in some circumstances will permit a design life as short as 10 years. It is commonly believed that codes such as the model code[6] aim for a design life of about 50 years.

It should be clear from the definition that reaching the end of the design life does not imply that the structure should only be fit for demolition; merely that the future cost of maintaining it in a fully functional state is likely to increase beyond that considered appropriate during its design life. A judgment would then have to be made as to whether the likely future maintenance costs were economically justified or whether demolition and rebuilding were more appropriate.

As far as this guide is concerned, further specific reference to design life is not made; the recommendations are aimed at ensuring that the design life implicit in the model code[6] (whatever that may be) is obtained.

De Sitter recently proposed his law of fives[14] (Fig. 8.1). This may be outlined as follows. The decline and fall of an unsatisfactory structure may be divided into four phases.

(*a*) Phase A: design and construction. The seeds of unsatisfactory performance are sown here, possibly due to bad design and material specification or poor workmanship.

(*b*) Phase B: pre-corrosion phase. Corrosion has yet to start, but carbonation or chlorides are penetrating inwards towards the steel more rapidly than is desirable. Remedial action could be taken if the problem is identified. This might, for example, consist of applying a carefully selected surface coating.

(*c*) Phase C: local active corrosion. Corrosion has started at some points and local spalling and rust staining become visible. Repair and maintenance will be necessary.

(*d*) Phase D: generalized corrosion. If repair and maintenance are not carried out, the structure will reach the state where major repairs are necessary, possibly including replacement of complete members.

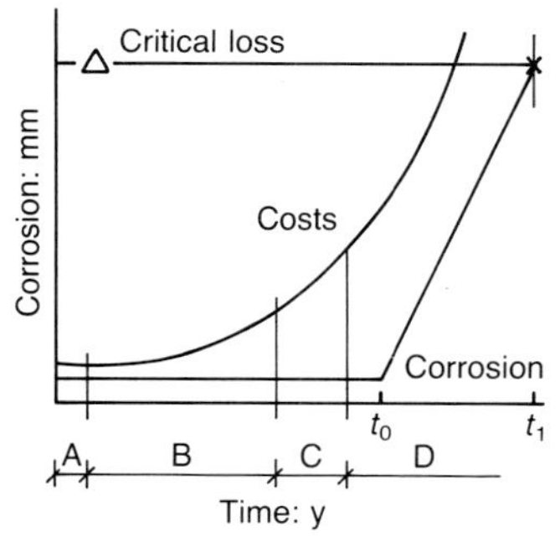

Fig. 8.1. The law of fives;[14] t_0 *marks the onset of generalized corrosion;* t_1 *marks the end of the service life*

De Sitter's contention is that $1 spent in getting the structure designed and built correctly in phase A is as effective as $5 spent in phase B, $25 in phase C or $125 in phase D.

It is not necessary to argue whether the rule of fives is absolutely correct or whether it should be a rule of fours or even threes; it remains clear that the most cost-effective way of ensuring an adequate life is to get the structure right in the first place. The objective of this guide is to help designers and constructors to achieve this.

A further point is worth emphasizing here in order to put all the design rules into perspective: the factors which have by far the greatest influence on the durability of concrete structures are adequate compaction of the concrete and good curing. If these are not achieved, the efforts of the designer are almost totally wasted. It follows from this, however, that anything the designer does to make the structure easier to concrete will pay handsome dividends in improved durability.

Curing and compaction are particularly important for the concrete in the surface layer. It is this layer which is directly in contact with the environment: it protects the steel and, in the case of chemical attack, it is most at risk. Unfortunately, this is the concrete which is more likely to be poorly compacted and poorly cured. For these reasons, the recommendations in section 10.5 are of the most fundamental importance in ensuring durable structures.

A sound understanding of the phenomena related to the deterioration of structural concrete is the best foundation for achieving durable structures. Part I of this guide gives sufficient information on the various mechanisms involved for this to be obtained. In part II the aim is to give more directly useful practical advice on specific issues.

9. Classification of environmental exposure

9.1. Definition of exposure classes

The Comité Européen de Normalisation (CEN) has recently submitted a draft European standard[11] on concrete, covering performance, production, placing and compliance criteria, for a preliminary vote by member countries. This draft includes a comprehensive categorization of exposure classes (Table 9.1) which may be compared with Table 9.2 which covers the special problems relating to reinforcement. The CEN definitions are more detailed than those in the model code[6] (chapter 7).

9.2. Assessment of chemical attack on concrete

A quantification of the degree of aggressivity of the environment is useful, although it may represent a simplification in cases where combined attacks

Exposure class		Environmental conditions
1		Dry environment, e.g. — interior of buildings for normal habitation or offices — exterior components not exposed to wind and weather or soil or water — localities with higher relative humidity only for a short period of the year (e.g. >60% RH for less than 3 months per year)
2	a	Humid environment without frost,* e.g. — interior of buildings where humidity is high — exterior components exposed to wind and weather but not exposed to frost — components in non-aggressive soil and/or water not exposed to frost
	b	Humid environment with frost,* e.g. — exterior components exposed to wind and weather or non-aggressive soil and/or water and frost
3		Humid environment with frost* and de-icing agents, e.g. — exterior components exposed to wind and weather or non-aggressive soil and/or water and frost and de-icing chemicals
4	a	Sea-water environment, e.g. — components in splash zone or submerged in sea water with one face exposed to air — components in saturated salt air (direct coast area)
	b	Sea-water environment with frost,* e.g. — components in splash zone or submerged in sea water with one face exposed to air — components in saturated salt air (direct coast area)
The following classes may occur alone or in combination with the above classes		
5†	a	Slightly aggressive chemical environment (gas, liquid or solid)
	b	Moderately aggressive chemical environment (gas, liquid or solid)
	c	Highly aggressive chemical environment (gas, liquid or solid)

* Under moderate European conditions.
† See ISO classification of chemically aggressive environmental conditions affecting concrete. The ISO standard is still to be established. See also Table 9.3.

Table 9.1. Exposure classes for concrete related to environmental conditions[11]

Table 9.2. Exposure classes for reinforcement related to environmental conditions

Exposure class		Environmental conditions
1		Dry environment: generally dry localities of fairly constant humidity when the relative humidity only infrequently exceeds 70%, e.g. interiors of buildings for normal habitation or offices
2	a	Environments with infrequent major variations in relative humidity, giving only occasional risk of condensation
	b	Environments with frequent major variations in humidity, giving frequent risks of condensation
3		Humid environment with frost* and de-icing agents, e.g. exterior components exposed to wind and weather or non-aggressive soil and/or water and frost and de-icing chemicals
4		Sea-water environment, e.g. — components in splash zone or submerged in sea water with one face exposed to air — components in saturated salt air (direct coast area)

* Under moderate European conditions.

Table 9.3. Assessment of the degree of chemical attack of concrete by waters and soils containing aggressive agents (from ref. 16, after ref. 14)

	Exposure class* 5a	Exposure class* 5b	Exposure class* 5c	
Type of attack	Weak attack	Moderate attack	Strong attack	Very strong attack
Water				
pH value	6·5–5·5	5·5–4·5	4·5–4·0	<4·0
Aggressive CO_2: mg CO_2/l	15–30	30–60	60–100	>100
Ammonium: mg NH_4^+/l	15–30	30–60	60–100	>100
Magnesium: mg Mg^{2+}/l	100–300	300–1500	1500–3000	>3000
Sulphate: mg SO_4^{2-}/l	200–600	600–3000	3000–6000	>6000
Soil				
Degree of acidity according to Baumann–Gully	>20	X†	X†	X†
Sulphate: mg SO_4^{2-}/kg of air-dry soil	2000–6000	6000–12 000	12 000	X†

* See Table 9.1.
† X = conditions of attack which are not found in practice.

occur. Cembureau[15] has produced recommendations that can be of value; Table 9.3 presents the assessment of the degree of chemical attack of concrete by water and soils containing aggressive agents.

10. Design, construction and maintenance

10.1. Handling the building process

A traditional building process is characterized by a specialized input from all the parties involved

(*a*) the owner (client) by defining his demands and wishes
(*b*) the designers (engineer and architect) by preparing design, specifications (including control schemes) and conditions
(*c*) the contractor, who will try to follow these intentions in his construction work; subcontractors are also commonly involved.

In this traditional picture, one of the important parties is not mentioned: the user of the structure (the building), who will normally be responsible for the maintenance of the structure during the period of use.

The influence of the above-mentioned parties on the quality of the final product can be seen from the quality circle of a building (Fig. 10.1). Any of the four parties may — by their actions or lack of attention — contribute to an unsatisfactory state of durability of the structure. Also, interactions between two parties may cause faults which can have an adverse effect on the durability. It is well known that in cases of premature deterioration, any of the parties may, and usually will, blame the other parties for the poor results. Such an attitude is not very productive — and in most cases it is basically wrong. All parties are normally equally responsible for shortcomings of this kind, and a contribution from all of them is necessary if the outcome of a building process is to lead to lasting structures and be to the satisfaction of all involved.

Furthermore, it is important to realize that the concrete durability problems experienced in the past can only be avoided in the future if adequate and co-ordinated efforts are imposed on all phases of the process of defining, planning, building and using the project until the end of its expected lifetime. The goal for such efforts should be to select methods and perform actions throughout the construction process which will result in optimized overall

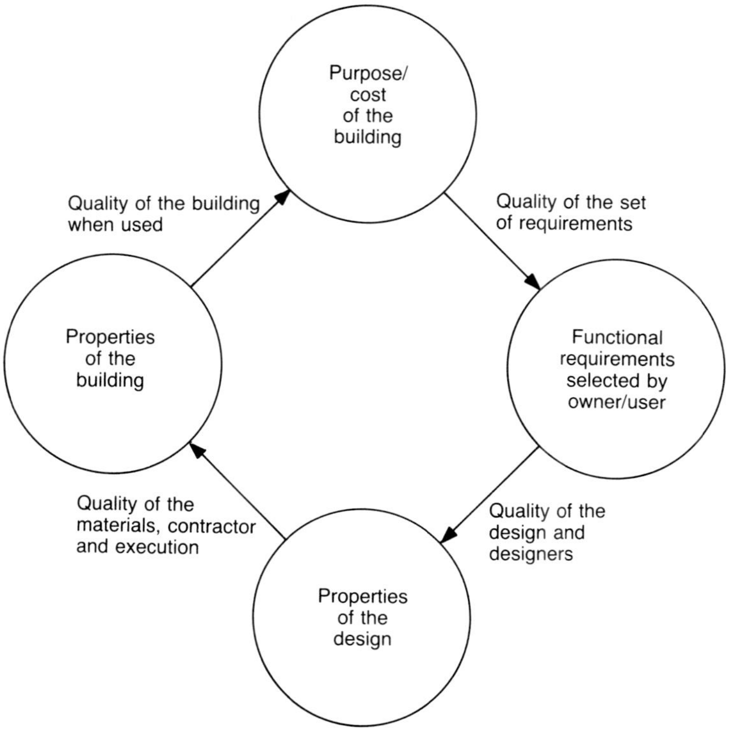

Fig. 10.1. The quality circle for a building

Table 10.1. Interactions between phases and parties involved in building and using concrete structures[17]

Phase	Party involved	Interactions
Definition	Client (owner)	Define use of building
Planning	Consultant	* Environmental conditions * Service period
Design	Consultant Architect Engineer Client	* Construction material * Structural concept * Important details * Construction process Codes of practice Loads * Environmental impact Safety during planned lifetime Specification Construction drawings Technical report
Approval of design	Client Authorities	Standard regulations Codes of practice
Construction	Contractor Consultant Client	Construction programme * Concrete constituents * Testing, choice * Concrete mix design * Trial concreting (structural conditions) * Testing * Execution Quality assurance system Quality statement record
Preliminary handing-over	Contractor Consultant Client	Certificate of substantive completion
Maintenance period	Contractor	* Remedial works * Maintenance
Final handing-over	Contractor Consultant Client/user	Final certificate of completion
Period of use	User Maintenance consultant Owner User Specialized consultant	Initial inspection Maintenance manual Data report Routine inspections * Preventive maintenance When deemed necessary, special investigation * Maintenance and renovation * Maintenance and repair

* Principal interactions affecting durability.

costs for the creation of the project and in proper functioning during the period of use.

Table 10.1 gives one example of possible interactions between phases and parties involved in the process of creating structures, showing the responsibilities of the different parties involved and the sequence of actions and decisions they are expected to perform. The transition between different phases and the direct interaction between the different parties in the process are especially sensitive to shortcomings in transmitted and received information. In the example, various means of recording and transmitting information are noted.

The planning and design will be based not only on the intended use of the structure, but also on the environmental conditions and the planned service period. A technical report outlining the basis for and the results of the design process will give the client a clear picture of the project — and its limitations.

Trial concreting under structural conditions and subsequent relevant durability testing is an important part of the preparatory work. The result of the construction process should always be described in an official quality statement record.

After usual contractual handing-over, it is important to have the maintenance started by an initial inspection and the preparation of a maintenance manual. Then later, systematically recorded routine inspections will form the basis of decisions regarding necessary maintenance work. When unusual or serious problems are disclosed, specialist investigations shall be the basis for decisions on extraordinary maintenance works.

Quality statement records and inspection records can provide useful knowledge and experience as a basis for future practice and decisions, and the client's and the user's understanding would benefit from a more comprehensive presentation of problems and results.

The designer's work and responsibility will become more precise. He will realize that in many respects he will have to extend his knowledge or look for specialist assistance. He will further recognize the need for adequate education of specialized concrete materials engineers and for an improvement in the education structure, where there seems to be some disharmony between the highly developed computation methods and an adequate knowledge of structural detailing.

Also, the contractor and his staff will benefit from well-defined terms and from a more thorough recording of the results obtained. The challenge to achieve good and uniform results by fulfilling prescribed criteria will become more marked if the results are not only observed but also properly recorded for future use. A successful improvement of the durability of concrete structures can doubtless be achieved if the codes of practice — and hence the model code — reflect the intention to include durability (over a planned service period) in the design basis.

It is believed by the CEB General Task Group on Durability and Service Life of Concrete Structures that general considerations along these lines may be useful in an attempt to preserve new concrete structures for a sufficiently distant future.

The following sections treat different practical aspects of this scheme.

10.2. Workmanship

A high proportion of analysed structural and functional defects can be attributed to infringement of acknowledged rules of design and construction, to insufficient training and expertise of personnel, or to simple lack of attention, and should therefore be avoided.

10.2.1. Motivation, information and education

The single most important element in preserving and improving the quality of structures and their performance is efficient continuing education, where

Fig. 10.2. Balconies in prefabricated concrete

Fig. 10.3 (below left). Protruding pillars on a conference building

Fig. 10.4 (below right). Deteriorated pillar

new theories, new technologies and experience gained can be spread to a sufficiently large number of people involved in the design, construction and upkeep of building structures.

Supplying relevant and current information to the persons involved is acknowledged as the best means of motivating and involving them in the work, thereby reducing faults and errors due to neglect or lack of knowledge.

In the building and construction sector a quality assurance manual is a valuable document which helps to keep a clear overall view of activities and processes needed to create complicated structures, and especially to keep track of interdependence and timing during the building process. Furthermore, such a document is a valuable place in which to keep updated information on procedures and techniques.

However, not even the most strict control procedures can compensate for a lack of personal motivation to produce a good and reliable product.

More detailed recommendations, relevant to the main phases of the life of a structure, are now given.

10.3. Design and detailing

Structural design, comprising architectural concepts of layout together with engineering selection of structural form, determines the overall geometry of the structure, including the exposed parts (Figs 10.2–10.4). In local or microscale this in turn affects the type and intensity of possible deleterious interaction between the structure and its environment.

By following the tradition of focusing primarily on durability aspects of the material composition, the importance of the structural form in determining the long-term durability and performance of a structure may well be overlooked. In this respect, architectural designs well thought of from the point of view of required long service life may well differ considerably in aesthetic appearance from a large number of today's buildings and structures.

However, not only the general structural layout of exposed surfaces is of importance in determining the actual rate of attack of an aggressive environment. Often, small and simple details related to the design, execution and maintenance may tip the scale in deciding whether or not the structure will obtain longevity. Most of these details are covered in this section. The cases presented are only to be considered as simple examples of the general principles and are not intended to give a fully comprehensive covering of the topic.

One general conclusion which can be drawn from the following sections is that complexity causes trouble and that the more robust designs may result in the most durable structures.

10.3.1. Durability is about drainage: no water — no trouble

10.3.1.1. Drainage over concrete. Avoid conditions where water drains over concrete, or over joints and seals (Fig. 10.5). If water from rain, melting snow and ice, drainage outlets and so on is allowed to drain over concrete, water and dissolved aggressive agents such as chlorides may penetrate into the concrete, or the concrete may be washed out, endangering the concrete as well as the reinforcement.

Where watertight joints and seals are necessary, their long-term tightness

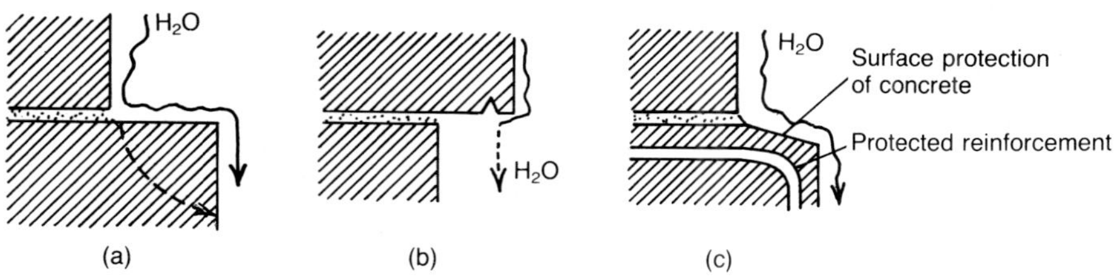

Fig. 10.5. Water draining over joints and seals: (a) this set-up should be avoided; (b) this arrangement is preferable; (c) some form of surface protection is another option

cannot be relied on, and possible consequences of their malfunction should be foreseen. This may require draining slopes on the top surface of supporting beams or columns and perhaps even special water protection or drainage of these zones, although such measures only come into use in the case of a joint malfunction. Where de-icing salts are used on bridges, parking decks or balconies, leaky joints may cause chloride corrosion of the otherwise fully protected supporting elements, resulting in serious local degradation with consequences in complete disproportion to the costs of avoiding the cause.

10.3.1.2. Standing water. Conditions where water can stand should be avoided (Figs 10.6 and 10.7). Exposed surfaces that need to be close to horizontal (e.g. parking decks, balconies, pavements and bridges) should be drained away from critical zones such as joints and seals, and the drainage should be correctly achieved and maintained (Fig. 10.8).

Smooth surfaces for façades shed water more easily than rough ones. However, surfaces with exposed aggregates are much less absorbent to water (see chapter 11).

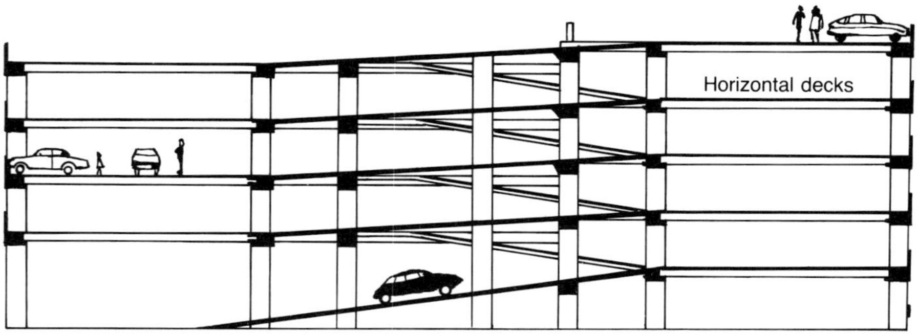

Fig. 10.6. Multi-storey car park, with horizontal decks where water can stand

Fig. 10.7 (below left). Deck of a multi-storey car park, showing water collecting

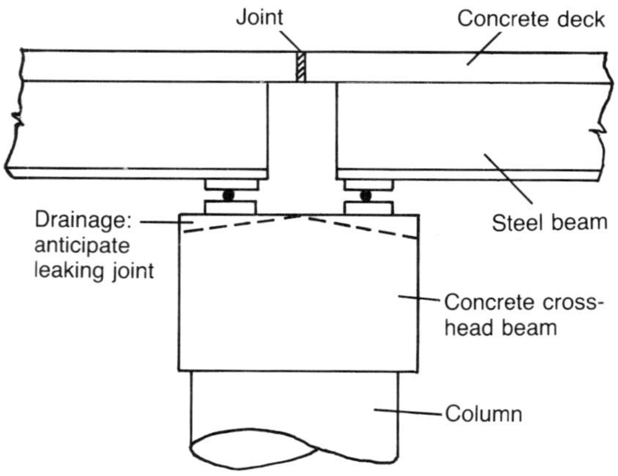

Fig. 10.8 (below right). Drainage at a joint

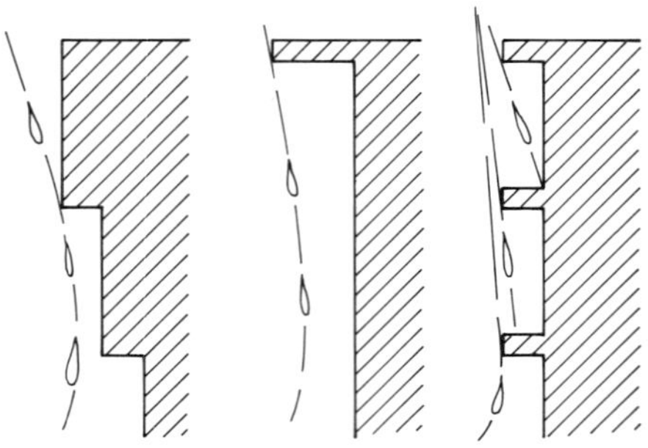

Fig. 10.9 (below). Protection of façades from rain

Fig. 10.10 (right). Column close to a roadside

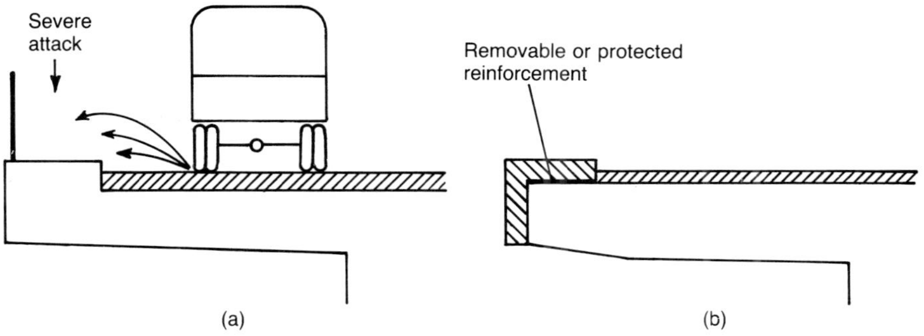

Fig. 10.11. Splashing: (a) a situation in which splashing is likely to occur; (b) protection added

Fig. 10.12. Freezing and bursting of concealed water

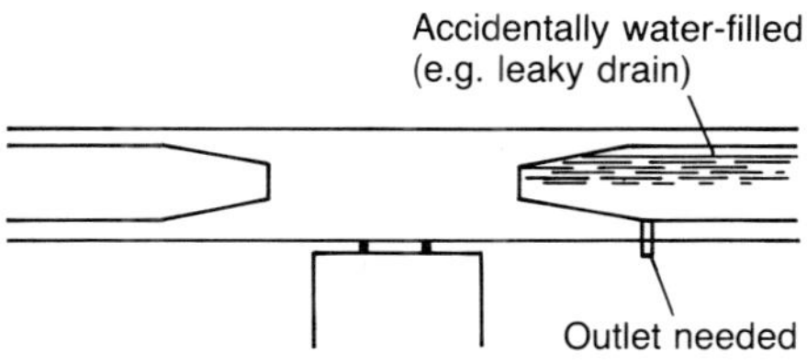

Fig. 10.13 (far right). Drainage of voids

10.3.1.3. Splashing. Surface areas subject to wetting or splashing should be reduced. Roofs with large eaves provide valuable protection of the façades against wetting from rain. Bands of balconies may have a similar effect (Fig. 10.9).

The economic building style where eaves of the roofs are left out altogether has probably caused the owners substantially larger sums for maintenance and repair due to excessive wetting and drying of the façade than was gained by the shortsighted initial savings in construction costs. This is valid not only for concrete structures, but also for masonry and timber.

Retaining walls and bridge piers close to traffic roads may profit from having a larger distance to the road than the minimum, as splash water and fog spraying caused by the traffic are reduced (Fig. 10.10). This is especially true if de-icing salts are used. Although construction costs for a bridge, for example, would increase with increased spans, this may well be an advantageous solution in the long run.

10.3.1.4. Protection against splashing. Surfaces where splashing is possible or where drainage is difficult should be protected (Fig. 10.11). In such cases a special structural protection such as a screen wall or an easily replaceable element may be provided. Surface coatings may also be valuable, provided that the correct penetration or diffusion characteristics are achieved regarding moisture, air and aggressive substances.

The watertight membrane often applied to bridge decks is an example of such special protection.

10.3.1.5. Drainage. It is necessary to ensure good drainage and ventilation. Water may accumulate in any void present in an exposed structure. This may increase moisture conditions and raise concentrations of dissolved aggressive substances in the surrounding concrete to critical levels. Deleterious effects may develop without being visible on the outside, giving rise to risks of malfunction and failure without warning.

If much water accumulates in such voids, freezing may cause sudden bursting of the surrounding structural concrete, causing partial or even total failure of the element (Fig. 10.12). Voids in slabs and the hollow space in the box girders should, therefore, always be safely drained and ventilated (Fig. 10.13). Preferably, they should also be inspectable (Fig. 10.14) (see section 10.6.2).

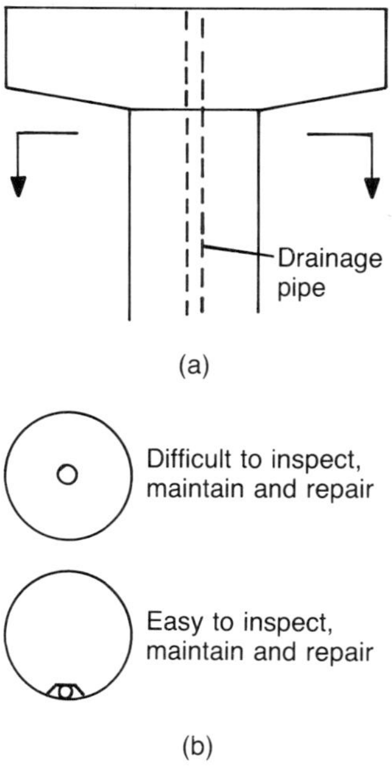

Fig. 10.14. Ease of maintenance of drainage pipes: (a) side view; (b) cross-section of two possible arrangements

10.3.2. Large cracks allow ingress of aggressive substances

Conditions that are likely to lead to large cracks should be avoided. Abrupt

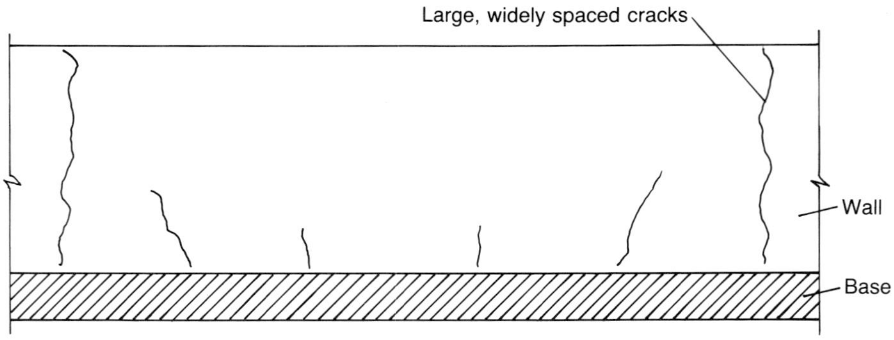

Fig. 10.15. Large local cracks

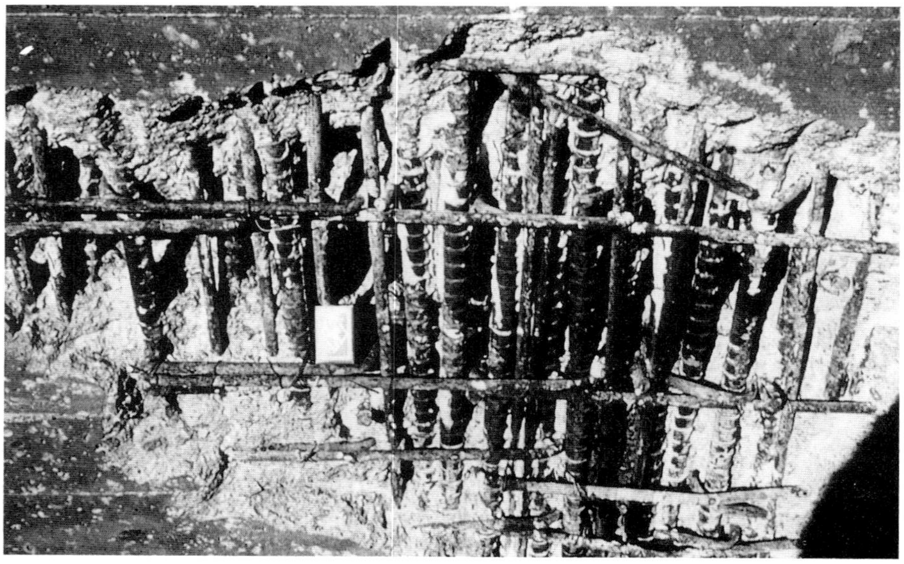

Fig. 10.16. Inappropriate concreting due to inappropriate reinforcement detailing

Fig. 10.17 (below). Detailing of reinforcement: (a) appropriate concreting and compaction are not ensured; (b) gaps are available for the insertion of a vibrator, and the bar spacings are sufficient for appropriate concreting and compaction. The dimensions of the cross-section should be enlarged if necessary

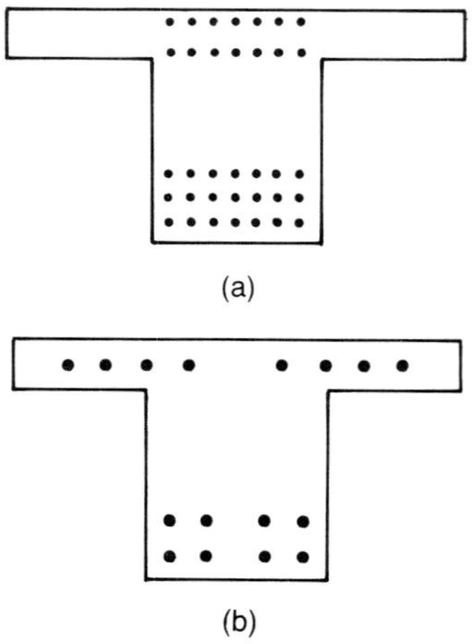

deviation of forces in a structure and abrupt changes in sections result in stress concentrations likely to cause cracks. The corresponding detailing of the reinforcement may in itself be crack-initiating, although it may distribute the cracking and reduce the crack widths.

Concentrated forces due to anchoring of prestressing tendons or due to reactions from supports create large local splitting forces which cause cracks if not dealt with by an appropriate reinforcement. Restraining forces due to, for example, differential settlement, shrinkage and temperature effects may also cause large, local cracks if not adequately foreseen in the design and reinforcement detailing (Fig. 10.15).

10.3.3. Spalling reveals bad reinforcement detailing

Although the reinforcement is hidden within the concrete of the finished structure, its detailing has considerable influence on the durability of the structure. Inappropriate reinforcement detailing may be revealed by early corrosion and spalling of cover initiated by large cracks, locally porous concrete or insufficient cover (Fig. 10.16). Care should be taken to ensure a detailing which takes durability aspects into account and to control the execution accordingly (Fig. 10.17).

Section 10.3.2 treats the influence of detailing on cracking, and section 10.5.1 considers the interaction between the reinforcement detailing and the execution. Further details are given in refs 7, 18 and 19.

10.4. Material composition

The ability of reinforced and prestressed concrete to withstand use and adverse environments depends to a large extent on the initial quality of the concrete and steel. This section is to be considered as a check-list to ensure that the important parameters have been considered in design.

10.4.1. Good concrete depends on good components

The potential of modern concretes to cope with even very adverse chemical environments, together with their (at times) extreme sensitivity to correct and careful handling, especially during hardening, makes it essential to evaluate the concrete mix carefully. This includes the chosen or available cement, together with the type, composition and grading of the available aggregates, the mixing water and possible admixtures. The single most decisive parameter in determining the permeability of the outer concrete layer is the W/C ratio, which should be low.

10.4.1.1. Cement. The characteristics of concrete regarding permeability, chemical binding capacity and resistance to aggressive agents depend considerably on the type of cement used. Blending agents in composite cements, especially pozzolana and slag, generally improve resistance against most of the chemical attacks but may increase curing sensitivity and decrease the resistance against frost and carbonation, especially if the concrete is insufficiently cured. It is therefore necessary to make a careful selection of the cement when specific requirements regarding concrete composition and environmental aggressivity must be met.

10.4.1.2. Aggregates. Alkali-reactive and non-frost-resistant aggregates are unsound and should be avoided. Aggressive substances such as chlorides and sulphates and organic and inorganic impurities such as humic acid, clay and other fine impurities must not be overlooked when evaluating the suitability of aggregates for concrete mixes.

Modern techniques of density separation of aggregates and means of selecting inert aggregates are available to help solve these problems.

10.4.1.3. Mixing water. Drinking water is usually acceptable as mixing water, but if in doubt it should be tested. The mixing water may be polluted with aggressive substances such as Cl^-, SO_4^{2-}, NO_3^- and alkalis (Na^+, K^+). These and other impurities may contribute unfavourably to the total content of aggressive agents mixed into the concrete.

10.4.1.4. Mineral additives. Mineral pozzolana added to the concrete mix reduces the development of hydration heat, may contribute positively to the strength development at later ages, and may improve the resistance to chemical attack considerably, but increases the curing sensitivity and may have negative effects on frost resistance. Special care should be taken when combining mineral additives with composite cements.

10.4.1.5. Admixtures. The chemical composition of admixtures (e.g. plasticizers, air entraining agents, accelerators and retarders) is often difficult to discover, but they may contain agents highly detrimental to the concrete or the reinforcement (ordinary and prestressed). For example, calcium chloride is a well-known and efficient accelerator, but when used in reinforced structures (ordinarily reinforced or prestressed) the consequences may be disastrous (see section 6.2.5).

10.4.2. Durable reinforcement depends on good concrete

The quality and thickness of concrete cover and the crack width should be such that adequate protection is provided against depassivation (carbonation, chloride contamination) and corrosion within the anticipated service life of the structure (see section 6.3.1). Of special concern are prestressing reinforcements, where special measures may be needed against the dangers of brittle failure caused by stress corrosion cracking or hydrogen embrittlement (see section 6.2.6).

10.5. Execution and curing

Investigations of the primary causes of premature deterioration of concrete structures, reinforced as well as prestressed, reveal nearly unanimously that apparently minor discrepancies that occurred during the execution phase and during the period immediately following were responsible in the majority

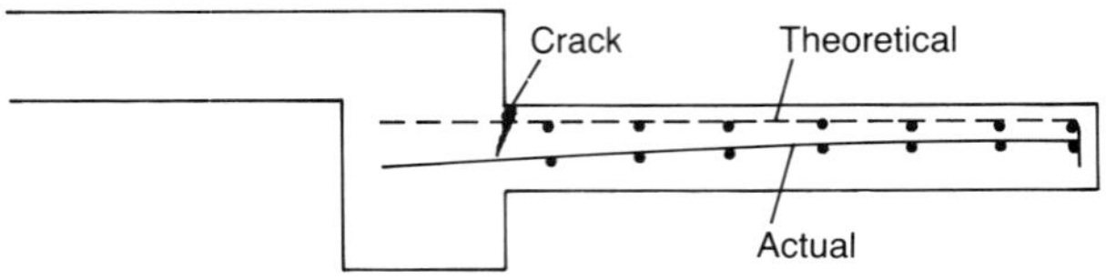

Fig. 10.18. Displacement of the reinforcement in a cantilever (balcony)

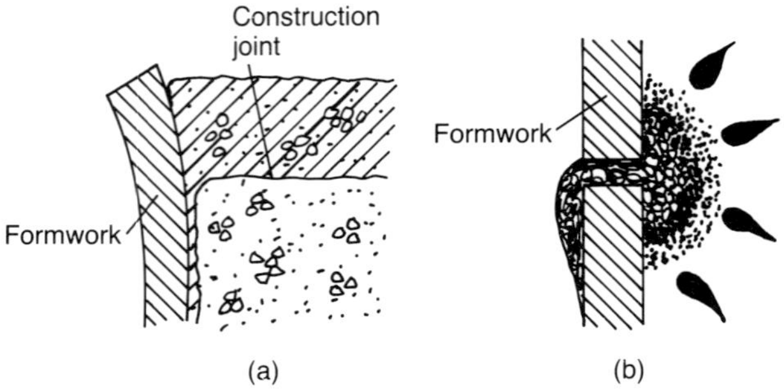

Fig. 10.19 (right). Formwork: (a) displacement of formwork; (b) leakage through formwork

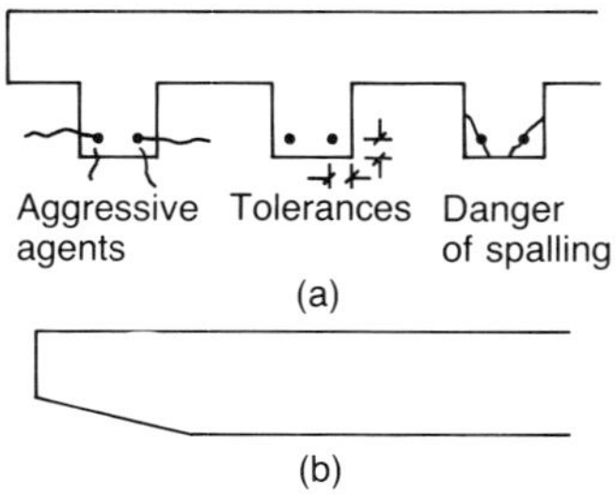

Fig. 10.20. Complex geometry should be avoided: (a) is liable to lead to all kinds of problems; (b) is better

of cases. This includes inadequate composition of concrete, poor concreting and insufficient curing. Numerous cases of damage are caused by too high a permeability and insufficient thickness of the concrete cover, the latter being perhaps the single most important factor determining the durability and service life of the entire structure.

10.5.1. Well-constructed structures will be durable

A structure which is easy to construct will be more likely to be constructed properly, and hence be durable.

Difficult details should be avoided. Reinforcement should be easy to place and compact concrete around. It should be fixed firmly in the form to avoid displacement, which may hamper proper placing and compaction of the concrete or may reduce the thickness of the cover (Fig. 10.18). Unreinforced sections or cuts should be avoided, as excessive cracks may develop.

Formwork must be stiff and well sealed. Leakage or displacements of the formwork may lead to porous or cracked concrete and to an unsightly surface (Fig. 10.19).

Complexity means trouble. Geometric form and reinforcement detailing should take constructability into account (Fig. 10.20). It is advisable to perform a constructability check before tendering on projects. These checks should be made by an experienced contractor.

Construction joints should be selected after careful consideration of the effects of reinforcement laps, bending and rebending of bars, anchoring of prestressed tendons and so on.

Prestressing systems require expertise, alertness and control. The measures needed to perform a reliable placing and stressing of prestressed tendons are well known and as such are trivial today. However, the process of grouting post-tensioned tendons in ducts seems to be a cause for concern. In a growing number of cases damage in the form of spalling and corrosion has been reported. This has also been reported for major prestressed structures with an age of 10–20 years or more.

The cause is ducts insufficiently filled with grout due to inappropriate grouting procedures, or cases where grouting has been forgotten altogether. For some reason water accumulates in this unintentional void, and although oxygen is scarce and corrosion thus extremely slow (except where ducts are ventilated via anchorages or unused grouting pipes), frost may eventually burst open the duct by spalling the concrete cover.

As such spalling may be hidden inside box girders or over water, undetected and accelerating corrosion may develop, jeopardizing the whole structure. Although grouting procedures have improved considerably over the years, this should be taken as a warning that execution processes resulting in hidden performance need especially good and reliable control.

10.5.2. Durable concrete depends on good curing — of good concrete

In chapter 8 it is emphasized that adequate compaction and good curing are the two factors having by far the greatest influence on the durability of concrete structures, and that this is of particular importance for the concrete in the surface layer. Curing of the concrete is part of the hardening process which ensures an optimal development of the fresh, newly cast concrete into a strong, impermeable, crack-free and durable hardened concrete.

During this initial stage of the life of the concrete, it is necessary

(*a*) to use an appropriate hardening process; casting must be planned such that the required strength at the time of form stripping is achieved
(*b*) to ensure against damage from drying; premature drying-out of the concrete surface should be avoided, as this may lead to large plastic shrinkage cracks
(*c*) to ensure against damage through early freezing; the concrete must not freeze until a required minimum degree of hardening has been achieved
(*d*) to ensure against damage from thermal stresses; differential movements due to thermal differences across the section or across a construction joint between hardened and newly cast concrete should not lead to cracks.

In recent years much valuable experience has been gained with the practical use of rational curing technologies. The increased sensitivity to too early drying out of some types of cement and concrete (composite and blended cements; chemical and mineral admixtures) has accentuated the need to develop simple and rational heat and moisture curing procedures.

A comprehensive presentation of such a curing technology is given in Appendix 1. It covers all phases of curing, from the calculation and planning to the control of the hardening process, including possible corrective measures to be enforced directly following observations during hardening.

Advice directly applicable in practice is now summarized.

10.5.2.1. Effect of temperature. The rate of hardening of the concrete is to a large extent determined by the temperature of the concrete. At 35°C the hardening is about twice as fast as at 20°C, and at 10°C the rate is about half that at 20°C.

For practical reasons, 20°C has therefore been chosen as a reference temperature, and through application of the temperature function H (Fig. 10.21) it is possible to compare hardening processes at other temperatures with an already known hardening process established at 20°C. The comparison is made by calculating the maturity M of the concrete, which is the equivalent age at 20°C.

Fig. 10.21 (below left). Temperature function defined for a thermally activated process. Relative velocity compared with the velocity at 20°C is given by $H = \exp\{E(\theta)/R \times [1/293 - 1/(273+\theta)]\}$, *where* R *is the gas constant. The empirical activation energy is given by* $E(\theta) = 33\,500 + 1470(20-\theta)$ *J/mol for* $\theta < 20°C$ *and* $E(\theta) = 33\,500$ *J/mol for* $\theta >= 20°C$

Fig. 10.22 (below right). Necessary pre-hardening time (maturity at 20°C) to obtain freezing strength of concrete due to self-desiccation, as a function of W/C ratio, showing data from a variety of studies. The dashed curves have been calculated from the equation $M \geq \tau_e/[-\ln(0{\cdot}86W/C)]^{1/\alpha}$

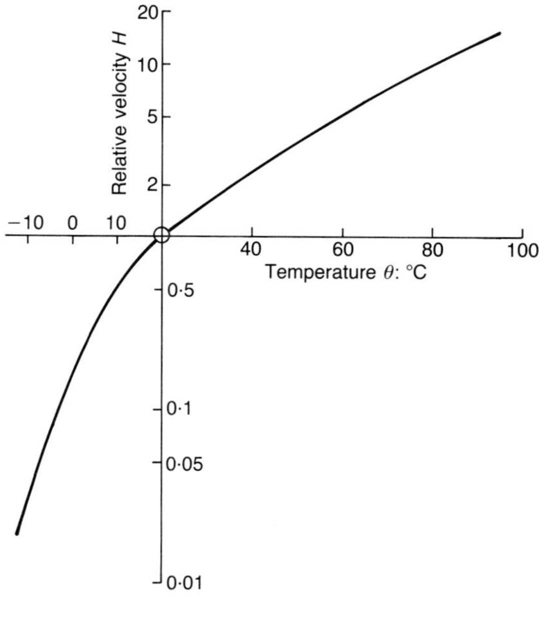

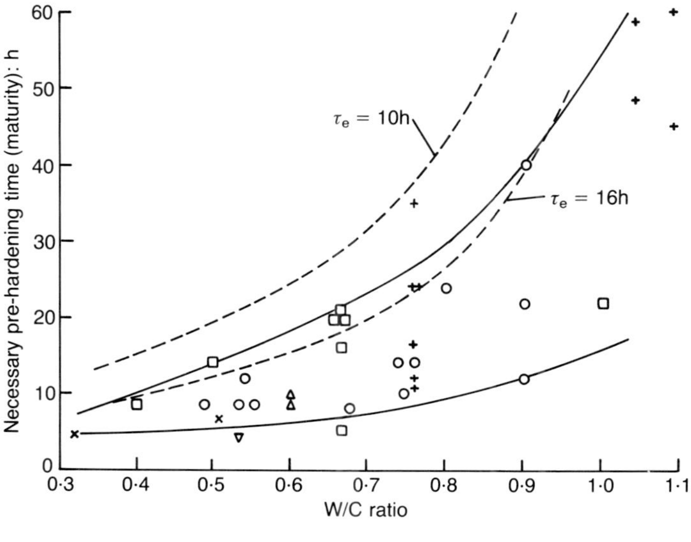

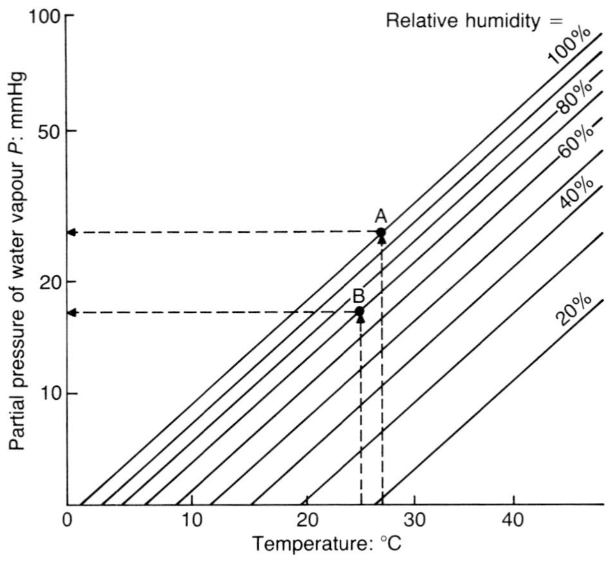

Fig. 10.23 (left). Partial pressure of water vapour as a function of temperature

Fig. 10.24 (below). Evaporation rate as a function of wind velocity and vapour pressure. It is assumed that the surface is wet until maturity at 10–20 h. The evaporation rate W *is given by* W = *(0·015+0·011*v*)*ΔP *kg/m²h, where* v *is the wind velocity*

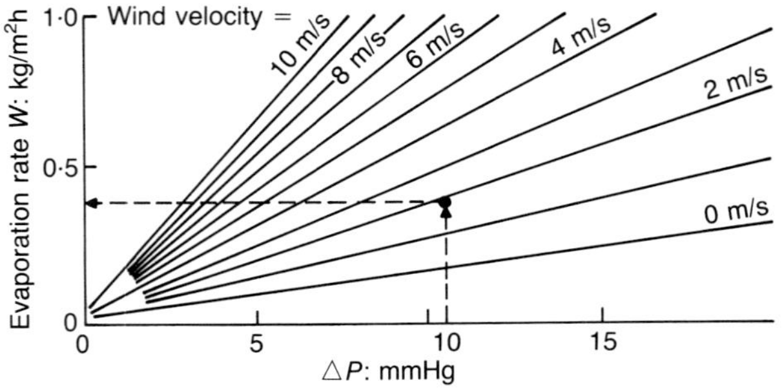

10.5.2.2. Prevention of premature freezing. If a hardened concrete freezes before a certain minimum degree of hardening has been achieved, the concrete may be damaged permanently. Figure 10.22 indicates the necessary pre-hardening time (i.e. maturity at 20°C) to obtain the minimum strength of concrete ≥5 MPa corresponding to sufficient hydration resulting in self-desiccation producing enough voids for freezing water to expand without damage to the concrete. The maturity is shown as a function of W/C ratio.

The pre-hardening time needed to achieve the required strength may be determined by calculation or testing.

10.5.2.3. Moisture curing. The evaporation of water from the concrete will take place as from a wet surface — provided sufficient water is led to the surface, e.g. by bleeding — until the reaction of the concrete corresponds to a maturity of 10–20 h. It is therefore particularly important to prevent excessive drying during the first 24 h after casting, if plastic shrinkage cracking is to be prevented. The actual quantity of water which may evaporate from a wet concrete surface can be estimated from Figs 10.23 and 10.24. The decisive factors in determining the rate of evaporation are the difference ΔP between the partial vapour pressure in the water layer on the surface of the concrete, and the partial vapour pressure in the ambient air.

The use of the diagrams can be illustrated by an example in which the temperature of the concrete — and the water — is 27°C and the relative humidity (RH) in the boundary layer is 100% (point A, Fig. 10.23). For the ambient air the temperature is 25°C and RH = 70% (point B). The difference in partial vapour pressure is then 27·0−16·5 = 10·5 mmHg. A wind velocity of 2 m/s is assumed, and so Fig. 10.24 gives an evaporation rate of 0·39 kg/m^{2}h.

It is not possible to give general rules for allowable rates of evaporation from concrete surfaces during initial hardening. These depend on the type of concrete, and especially on its tendency to bleed. For ordinary Portland cement concretes, the American Concrete Institute recommends that special precautions be taken if the rate of evaporation approaches 1·0 kg/m^{2}h. In the case of blended cements with little bleeding, a much lower limit is necessary.

Although bleeding is advantageous in reducing the risk of plastic shrinkage, it must not be forgotten that it also leads to porous concrete, especially near the surface, and thus bleeding should be reduced as much as possible within reason.

10.5.2.4. Heat curing. It is not possible to state exact limits to the temperature differences which are acceptable in hardening cross-sections, as they are dependent not only on concrete composition and strength characteristics, but also on the geometrical form of the hardening element. These limits depend also on the deformations and possible restraining forces due to the absolute temperature caused by hydration and the subsequent temperature drop to the level of ambient temperature.

According to experience, it is recommended to stay within the following limits for temperature stresses

(*a*) a maximum 20°C temperature difference over the cross-section during cooling after stripping

(*b*) a maximum 10–15°C difference across construction joints and for structures with greatly varying cross-sectional dimensions.

The heat balance to be controlled is sensitive to changes in the selected level of insulation. In practice it is often necessary to decide at short notice whether to strip formwork or whether possible additional or reduced insulation of a hardening cross-section has to be made. Figure 10.25 may assist in making this decision.

Good curing is needed to profit from a good concrete mix. Bad curing destroys an otherwise good concrete mix. Good curing cannot compensate for a bad concrete mix. All efforts to ensure an optimal heat and moisture curing may be in vain, if the initial quality of the concrete mix is inferior.

In practice, the temperature profiles can be calculated from the geometric data, the type of concrete, the type of curing conditions, and the ambient

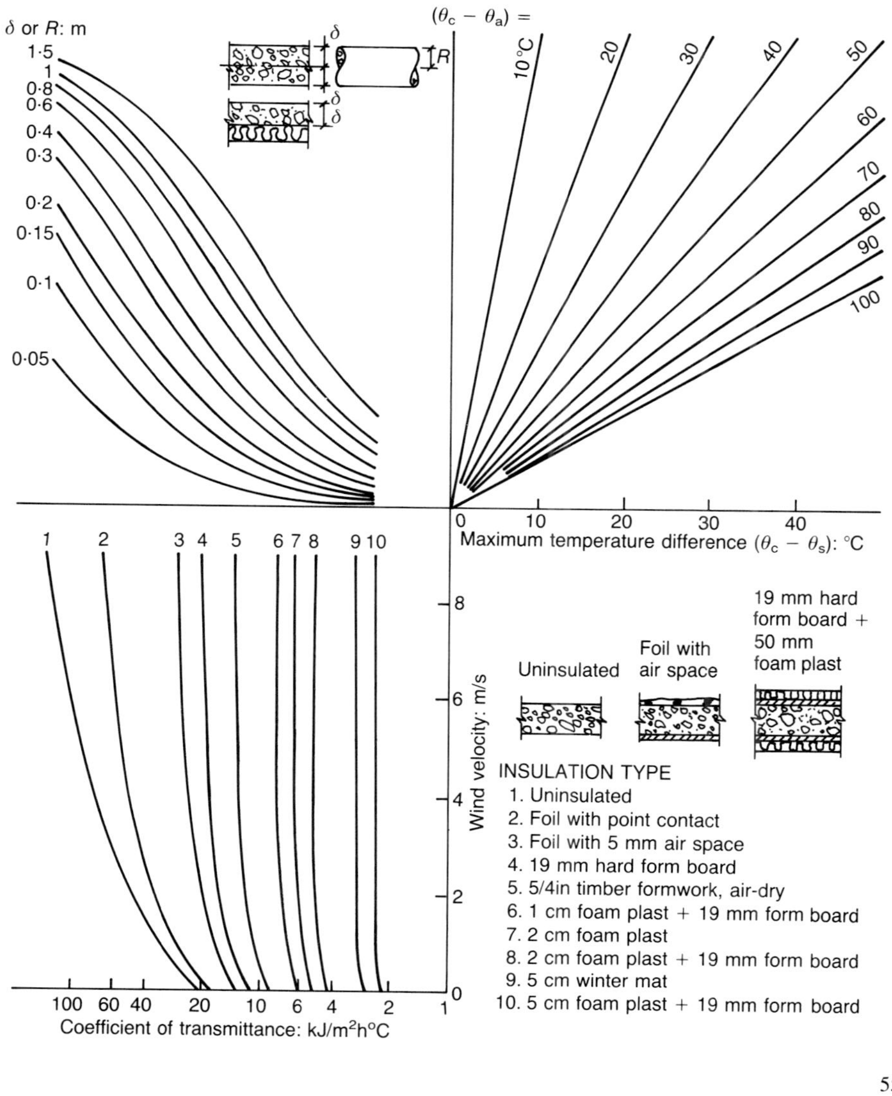

Fig. 10.25. Factors affecting stripping of formwork and insulation. The figure shows the estimation expression $(\theta_c - \theta_s)/(\theta_c - \theta_a) = B_i/(B_i + 2)$, *assuming standard concrete with thermal conductivity of 8·0 kJ/mh°C and density 2300–2400 kg/m³. Only contributions from conduction and convection are included; radiation, evaporation and condensation are not considered. The latter two can have a considerable effect on the coefficient of transmittance*

conditions. The resulting temperature differences should be compared with specified values, and necessary measures taken to satisfy requirements.

10.6. Service conditions

The actual safety and functional response of a structure in service depends partly on parameters chosen *a priori*, such as structural dimensioning, detailing, and choice of materials, and partly on specified or presumed parameters which in reality depend on the subsequent service conditions. These service conditions are unpredictable. This is also true to some extent for the ageing of materials. Hence, there is a need for regular inspection routines in order to maintain confidence in the structural integrity, performance and safety of the structure, and in order to assess the possible needs for maintenance, repair, strengthening or rehabilitation, as the case may require.

10.6.1. Service life is many things

The termination of the service life period is ideally the time when the structure becomes technically — or structurally — obsolete. However, in practice the usefulness of the structure may cease long before the technical or economical service life has been outlived. A sound structure may become functionally obsolete, e.g. allowable loads or required clearances may be increased. It is also possible to investigate the economics of upgrading the structure and thus extending the remaining service life.

When applying the service life concept in practice, the following types of ownership should be taken into account.

(*a*) The structure is owned and operated by one single responsible owner throughout its life. This may often be the state or large state-like organizations. Such structures may be, for example, power plants, nuclear plants, offshore structures or bridges.

(*b*) The structure is owned and operated by a multitude of successive private owners with relatively short horizons regarding economic involvement. This is the most usual case for ordinary dwellings, office buildings, and many factory-type structures operated under a private or capitalistic economy.

The large majority of structures are of the latter type, for which systematic inspection and maintenance cannot be fully relied on. In such cases, it is advisable to design and construct a robust structural skeleton and to rely on codes or specifications to ensure overall safety for the required lifetime.

For non-structural elements, finishes and installations, a shorter service life may be acceptable or even desirable, encouraging relatively frequent upgrading of the structure to meet the latest requirements of servicing, insulation, etc. Repair and modernization should thus be made easy to perform, and hidden installations or the like should be avoided.

10.6.2. Satisfactory service life requires inspection, maintenance and repair

Regular and systematic inspections should be performed in order to identify and quantify possible ongoing deterioration. Inspection constitutes an integral part of structural safety and serviceability by providing a link between the environmental conditions to which the structure is subjected and the manner in which it performs with time. The nature and frequency of the inspection procedures should be determined with this in mind.

In an advanced form the general strategy towards improved durability should incorporate systematic inspection routines for structures in service (including automated data recording and handling), decision models based on forecasting of the rate of degradation, and, as an important element, consideration of the economic consequences of taking either short-term or long-term remedial measures. To arrive at comparable figures for the economy of alternative solutions, present-day values of the future costs for maintenance, repair and eventual demolition and rebuilding must be sought.

These general procedures may be simplified when adapted to specific types of buildings, or to individual structures.

10.6.2.1. Accessibility for inspection and maintenance. When deciding on the final layout of a structure it is necessary to foresee which requirements must be fulfilled at the design stage in order to ensure reasonable conditions for inspection and maintenance in service. Buried elements (e.g. foundations and piles) or submerged elements are not usually readily inspectable during routine inspections. Only when malfunction puts these elements under suspicion may they be inspected, usually at high cost. Because they are so difficult to inspect, such elements should be constructed with the greatest care, incorporating a particularly high quality of material, and applying careful control.

10.6.2.2. Replaceability. The replaceability of particularly exposed elements with known short service life should be ensured. In many cases durability failures are the consequence of failure or malfunction of elements associated with the concrete structure, such as joints, bearings, drainage or the breakdown of waterproofing.

10.6.2.3. Prevention is better than cure. Preventive maintenance covers remedial work necessary to prevent expected deterioration or the development of defects. Whenever possible, the work should be done promptly — as soon as any incipient defects or conditions which may lead to defects are detected. Cleaning of the drainage system is perhaps the simplest example of preventive maintenance.

10.6.2.4. The decision not to repair. The assessment of a damaged structure may well lead to the conclusion that repair is too costly. In the case of a well-organized system of assessment and rating of structures this decision does not usually lead to immediate demolition, as the assessment routine should give ample warning before an unacceptable state has been reached.

There exists no clear strategy as to what technical and administrative measures to take when deciding on the consequences for use, inspection and maintenance once the decision of non-repair has been taken. The decision leads into an important but still gray area. The following questions should be considered.

(*a*) What should be looked for?
(*b*) How will the structure ultimately fail?
(*c*) Will there be any warning?
(*d*) Can an inspection procedure be devised that can serve safely as an early warning system?
(*e*) Should temporary maintenance work be performed with the aim of prolonging the replacement?
(*f*) How can an eye be kept on a condemned structure still in use?

One type of ordinary structure which may seem especially costly to repair is prestressed structures, when the deterioration directly involves the prestressing tendons, anchorages and couplers, or when the prestressed zone of the concrete is deteriorated and calls for replacement.

11. Weathering and discolouring

This chapter is based mainly on ref. 20. Reference 21 also contains useful information. The aim of the chapter is to explain the causes of the changing appearance of concrete surfaces and to give practical recommendations on how to prevent or limit these alterations.

Three phenomena change the original appearance of architectural concrete

(*a*) efflorescence, which is due to the capillary transport of lime towards the surface; it has no serious consequences, because of its temporary nature

(*b*) biological growth, which often adds to the unsightliness of concrete, and is usually mistaken for dust and dirt deposits; its main unfavourable effect is to keep the surface moist

(*c*) pollution, which is continuous and aggravates the situation.

Pollution in particular is treated in this chapter. The principal causes of pollution, the influencing factors and protective measures will be discussed.

11.1. Lime efflorescence

Due to the hydration of Portland cement, about 0·25 kg of slaked lime ($Ca(OH)_2$) is formed from each 1 kg of cement. Depending on concrete compactness, the time of demoulding and the climatic conditions, the dissolved lime moves to the surface and is transformed into carbonate due to the carbon dioxide present in the air. Efflorescences are activated by low concrete compactness, early demoulding and a dry and warm climate following a humid and cool period (Fig. 11.1).

Depending on the acidity of the rain, the lime dissolves without consequences for durability.

It is recommended to brush irregular and local efflorescence (e.g. stalactites near to cracks) as soon as possible, preferably before carbonation occurs.

Fig. 11.1. Lime efflorescences at the surface of a chimney wall in which cracks are present due to thermal gradients

Fig. 11.2 (below). Dust deposition on tall façades: in region A the wind velocity is high, and little deposition occurs; removal of dust may even occur; at B deposition is accelerated by the turbulence effect; at C deposition is increased due to traffic

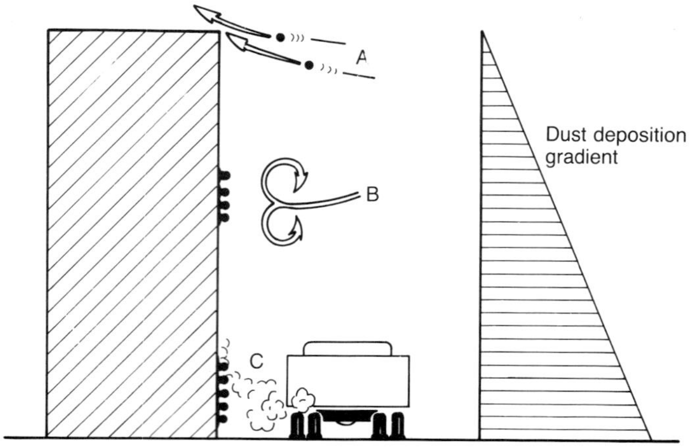

After carbonation, the efflorescences can only be removed with acid water followed by a thorough rinsing.

11.2. Biological growth

Concrete surfaces often provide the right conditions for the establishment of biological growths, but these are by no means always unsightly. Areas of algae or decaying lichens on concrete can be ugly; where attractive lichens occur on clean surfaces, however, they are often quite acceptable but usually go unnoticed.

Green or dark coloured algae will grow on most concrete that remains damp. Although some algae are known which can live on alkaline surfaces, reduction of the surface pH seems to speed colonization.

The full environmental factors governing the establishment of biological growths on building materials are only just beginning to be studied.

Many surfaces which appear to be dirty may be found on examination to have more biological contamination than mineral deposits, suggesting that an efficient way of including a long-lasting biocide in the surfaces would improve their appearance.

11.3. Pollution

11.3.1. Causes

11.3.1.1. Air pollution. The dust in the air is transported and deposited by the wind. Dust can be subdivided into

(*a*) fine dust (0·01 – 1 μm) which is in suspension in the air; it adheres to rough surfaces and has a great covering capacity due to a high surface : mass ratio
(*b*) coarse dust (1 μm – 1 mm), which is mostly of mineral origin; it has a small covering capacity.

Dust adheres less well to fast-drying surfaces than to surfaces which stay humid over a long period.

The wind influences dust deposition in two ways.

(*a*) Its velocity increases with height; the deposition of dust will be greater on the lower side of buildings, and this effect is intensified by the dust raised by traffic (Fig. 11.2).
(*b*) Near to an obstacle, the air stream is led away; the form of the stream pattern (laminar or turbulent) depends on the wind velocity. This stream pattern has a great influence on the dust deposition (Fig. 11.3).

11.3.1.2. Washing out by rain and trickling down. Pelting rain occurs due to the action of wind on rain. In northern Europe façades orientated between south and west are most exposed to pelting rain. They catch a mean of 40–50 l/m^2 of water per year.

The direction of the falling raindrops near to the exposed façade depends on the air stream at the different levels (Fig. 11.4). Wind velocity increases with height, and turbulence appears in the lower parts of the building. Tests have shown that maximum flow at a given level is not situated at the surface of the wall, but at a distance of 2·5–13·5 cm away, due to turbulence along the façade.

Pelting rain is often insufficient to wash out the dust and to clean the wall,

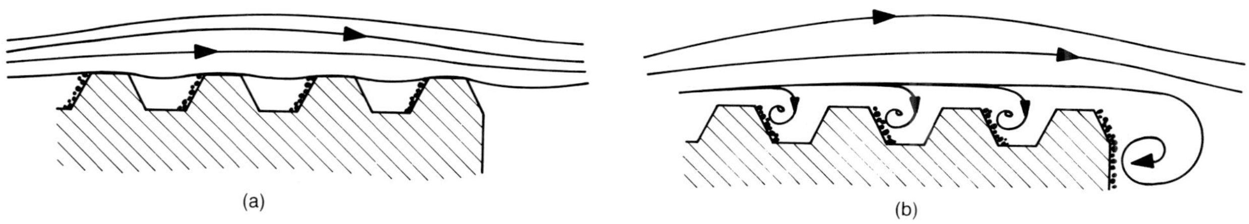

Fig. 11.3. Effect of wind stream type on dust deposition: (a) in a gentle wind, a laminar stream produces deposition on surfaces 'against the wind'; (b) in a heavy wind, a turbulent stream deposits dust on surfaces 'under the wind'

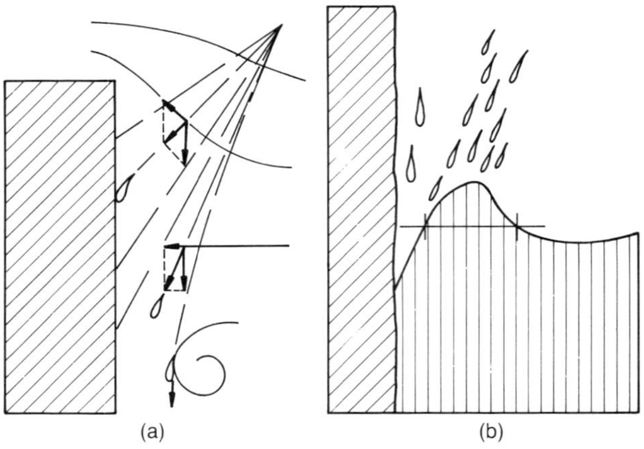

Fig. 11.4 (above). Rainfall near vertical surfaces: (a) the inclination of the raindrops varies with height; (b) the vertically shaded area shows the distribution of maximum rain flow with distance from the wall

Fig. 11.5 (right). Washing out and dust distribution

especially in the lower parts of the wall and for orientations other than between south and west.

The trickling down of rain is the main reason for pollution effects, because it sweeps away the uniformly deposited dust to redeposit it in a particular pattern (Fig. 11.5).

Horizontal or only slightly inclined surfaces will catch more rain than other types of surfaces. From this it follows that such surfaces are most subjected to the washing out effect by rain, especially for moderate or low rain intensities (north and east orientations).

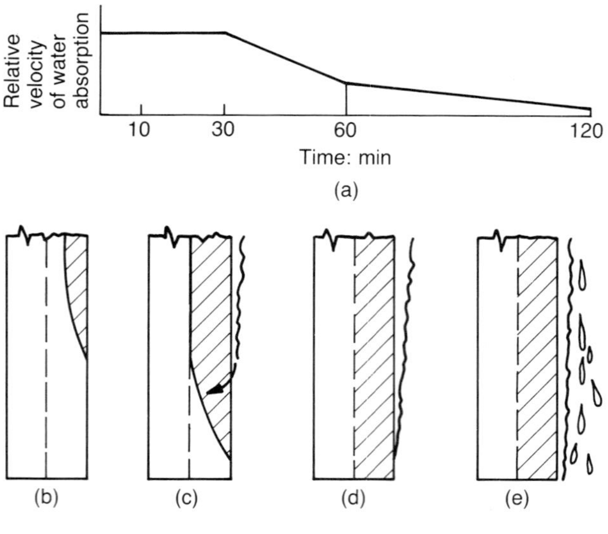

Fig. 11.6. Trickling down process: (a) relative velocity of water absorption with time; (b) absorption; (c) start of trickling down at the saturated part of the layer; (d) totally saturated layer and rain wash; (e) excess conditions and free drops of water

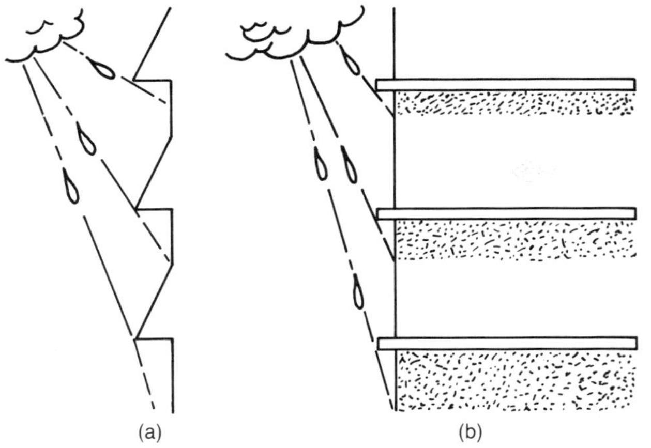

Fig. 11.7 (above). Two types of projecting edge, both protecting against wash-out

Fig. 11.8 (right). Example of a building on which projecting edges protect against wash-out

Fig. 11.9 (below left). Horizontal edges producing different types of trickle-down behaviour: (a)–(d) various types of edge (side views); (e) flow pattern on (b); (f) effect produced, as viewed from front

Fig. 11.10 (below right). Example of a building on which horizontal edges have produced trickle-down effects

11.3.2. The influence of the façade

11.3.2.1. Water absorption at the surface. As the effect of pelting rain is dependent on height, the trickling down starts at the top (Fig. 11.6). At the lower, non-saturated levels the water penetrates into the surface layer until saturation is reached. When the stage of continuous trickling is reached, some of the water falls directly on the ground. For normal rain on fairly porous surfaces, the dripping water seldom reaches the lower levels of the building.

11.3.2.2. Shape. Every withdrawn or projecting edge on the surface of the façade forms a protection against rain for the part beneath, from which dust is not washed out (Figs 11.7 and 11.8). Avoid even small interruptions in porches, especially for well washed façades.

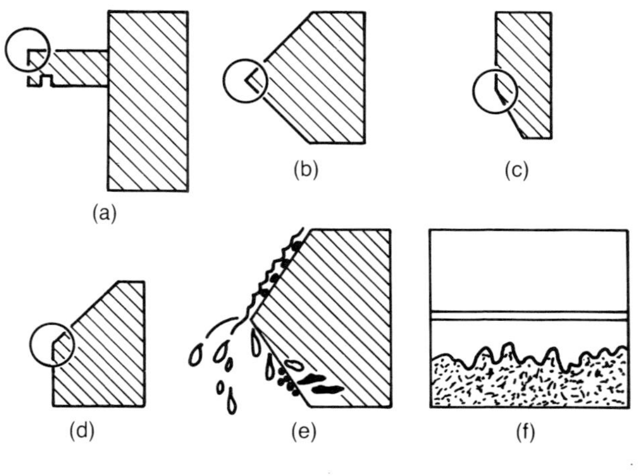

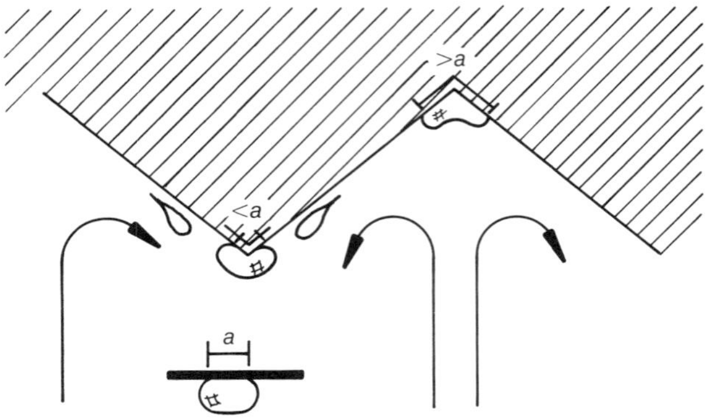

Fig. 11.11 (above). Trickling down is speeded up at protruding vertical edges and slowed at recessed edges

Fig. 11.12 (right). Example of trickle-down effects on vertical edges

Every horizontal edge is the boundary between two planes with different slopes, which have a different degree of moistening and trickling down behaviour (Figs 11.9 and 11.10). Every horizontal strip on which the water is evacuated at the outer side needs a front plane which can be totally washed. Its height should generally be limited. All horizontal strips with a certain width are favourable places for dust deposition. It is recommended that the profile be designed in such a way that rain is drained away at the inner side of the façade.

With regard to vertical edges, trickling down is accelerated at projecting edges and retarded at withdrawn edges (Figs 11.11 and 11.12). Special attention has to be paid to the crossing with horizontal edges, where trickling down is arbitrary.

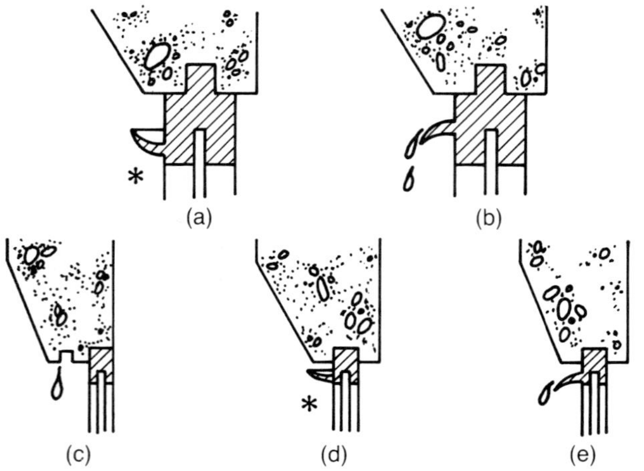

Fig. 11.13. Window designs; (a) and (d) collect water, and should be avoided

Fig. 11.14 (above left). Example of poor window design

Fig. 11.15 (above right). Good design. Deep, close grooves in the concrete surface regulate pollution. The vertical concrete surface under the window is free from running water. The water is drained and evacuated by a deep groove bordering the element

Fig. 11.16 (right). Poor design. The frame effect wanted by the architect is spoilt by irregular dust deposition due to the effects caused by horizontal surfaces and varying vertical planes

12. Measures against specific deterioration mechanisms

This chapter summarizes practical ways of coping with the specific deterioration mechanisms described in part I. It clarifies how the various recognized processes of degradation may be avoided or mitigated by profiting from a knowledge of the influencing parameters.

In practice a multitude of coinciding aggressive factors of varying intensity are present, thus seriously complicating the task of making the right decisions when deciding on materials, techniques and procedures influencing the service life of structures. A first step towards handling the complexity of actual environments is given in chapter 13.

The following recommendations are related to established and proven materials and technologies. Before using new materials (cements, blending agents, additives, admixtures, aggregates and reinforcing steel) or new technologies, the consequences with respect to durability must be checked. Figure 12.1 gives an example of the aspects to be checked with respect to the risk of corrosion of the reinforcement. Similar schemes should be followed with respect to other possible deterioration mechanisms.

12.1. Protection of concrete

In Table 12.1, limiting values for influencing parameters on the durability of concrete subjected to various environmental exposure classes (see Table 9.1) are given.

Concrete with a W/C ratio greater than 0·6 should not be applied for structural purposes.

Frost resistance may be achieved by means other than by air entrainment, e.g. by a low W/C ratio, where the value depends on the environment and the type of concrete constituents.

The recommended mix proportions given in Table 12.1 would normally ensure satisfactory durability, but if it can be proven by careful testing and control that the same values of the main quality parameters can be achieved with a revised mix (e.g. with a lower cement content) this may be acceptable (see section 6.3.3).

12.1.1. Protection against physical and mechanical action

12.1.1.1. Plastic shrinkage and settlement cracking. Parameters influencing the risk of cracking are dealt with in section 3.1.2. Supplementary comments are given here.

The most important parameters ensuring robustness in an otherwise well-

Fig. 12.1. Aspects to be checked in relation to corrosion of reinforcement

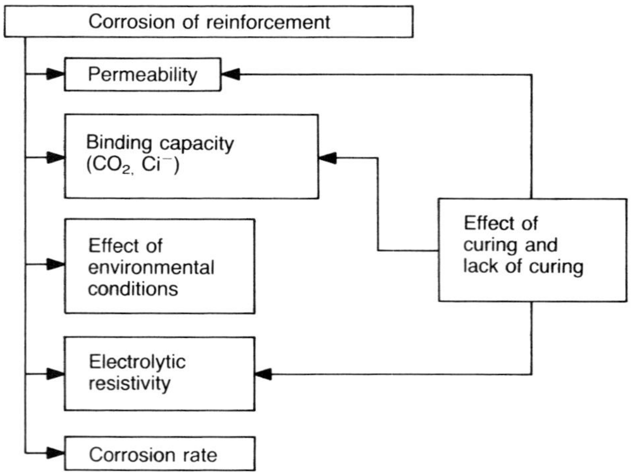

designed structure made with good initial quality concrete are proper and adequate heat and moisture curing.

There is a risk of plastic shrinkage cracking developing in a green concrete whenever the rate of evaporation from the concrete surface is greater than the rate at which water rises to the surface. Thus, especially in the presence of high air temperature, high wind, or low humidity, precautions must be taken to diminish the rate of evaporation. These precautions consist of moistening subgrade and forms, placing concrete at the lowest possible temperature, erecting windbreaks and sunshades, reducing time between placement of concrete and the start of curing, and minimizing evaporation by suitable means such as applying moisture (fog spraying), covering the surface with plastic or by curing membranes.

As composite and blended cements generally lead to a considerably reduced rate of bleeding of the concrete, they are more sensitive to the quality of

Table 12.1. Durability recommendations related to environmental exposure[11]

Requirement		Class of exposure according to Table 9.1								
		1	2a	2b	3	4a	4b	5a	5b	5c*
Strength class according to ISO 4012[22]	Plain concrete	≥C12/15								
	Reinforced concrete	≥C16/20	≥C20/25	≥C20/25	≥C20/25	≥C25/30	≥C25/30	≥C20/25	≥C25/30	≥C30/35
	Prestressed concrete	≥C20/25								
W/C ratio†	Plain concrete	—	≤0·70							
	Reinforced concrete	≤0·65	≤0·60	≤0·55	≤0·55	≤0·55	≤0·50	≤0·55	≤0·50	≤0·45
	Prestressed concrete	≤0·60	≤0·60							
Cement content† for maximum aggregate size between 16 and 32 mm: kg/m³	Plain concrete	≥150	≥180	≥180	≥180					
	Reinforced concrete	≥270	≥300	≥300	≥300	≥300	≥300	≥300	≥300	≥300
	Prestressed concrete	≥300	≥300	≥300	≥300					
Air content‡ according to ISO 4848[23] for maximum particle size of aggregates: %	≤32 mm	—	—	If risk that concrete will be saturated, as for class 3	≥4	—	If risk that concrete will be saturated, as for class 3	—	—	—
	≤16 mm				≥5					
	≤8 mm				≥6					
Water penetration according to ISO 7031:[24] mm		—	—	≤50	≤50	≤30	≤30	≤50	≤30	≤50
Additional requirements for aggregates		—	—	Frost resistant	Frost resistant	—	Frost resistant	—	—	—
Additional requirements for cement		—	—	—	—	—	—	—	—	Sulphate resistance§ when sulphate content in water >400 mg/kg, in soil >3000 mg/kg

* The concrete should be protected against direct contact with the aggressive medium by coating.

† Additions of type II (see clause 3, paragraph 12, of ref. 12) may possibly be taken into account, depending on the requirements applicable in the locality where the concrete is used.

‡ With spacing factor of air entraining agent = 0·20; however, no entrained air is required if the concrete, when tested according to ISO 4846,[25] satisfies the damage class 0 or 1.

§ The sulphate resistance of the cement shall be judged in accordance with the rules applicable in the locality where the concrete is used.

the workmanship during execution and curing, and a carefully planned and controlled moisture curing is needed.

In the case of plastic shrinkage and settlement cracks, a revibration of the concrete immediately after their formation can usually close them without damage to the concrete.

12.1.1.2. Cracking caused by loading and imposed deformations. Cracking is inevitable in concrete structures, reinforced as well as prestressed, and cracks do not *a priori* indicate undue lack of serviceability or durability provided that the crack widths do not become excessive. The width which can be accepted will depend on the function of the structure.

At the levels of stress currently used in reinforcement, cracking due to loading will not generally be sufficiently severe to lead to a reduction in durability or to seriously damage the appearance of a structure, provided that there is sufficient reinforcement to produce controlled cracking in those areas where tension is likely.

Excessive spacing of reinforcing bars will lead to wide, uncontrolled cracks between bars. The use of bars of large diameter relative to the cover may lead to the formation of cracks along the line of the bars. Bar spacing and bar diameters should therefore be limited.

Prestressing tendons create particularly high force concentrations in anchorage zones. Special reinforcement detailing has been developed to cope with this situation. For anchorages placed directly within the running cross-section away from the supports, such as dead-end anchorages or anchorages in construction joints, cracks can seldom be avoided, but may easily be controlled with careful detailing of the reinforcement in these areas.

A common practical way of dealing with cracks caused by imposed deformations is to design the structure such that the restraint is removed and the deformation allowed to occur freely. This can then be concentrated at points where measures can be taken to ensure that they do not cause problems. Expansion joints in buildings and bridges are examples of this approach dealing with external restraints.

A minimum reinforcement is particularly necessary in those parts of the structure where temperature, shrinkage or other actions can result in high tensile stress owing to restraints exerted on the imposed deformations. It should also be provided at construction joints subjected to tension.

Differential thermal cracking due to heat of hydration is dealt with in section 10.5.2.

12.1.1.3. Structural form and frost. As has been outlined in section 3.2, the most critical condition relating to frost resistance of concrete was found to be a water content close to saturation. Consequently, in design practice, the structural form should, as far as possible, be selected such that water saturation is prevented. Particularly susceptible to damage by frost are horizontal surfaces on which water tends to accumulate or vertical surfaces along which water flows due to incorrect drainage (Fig. 12.2).

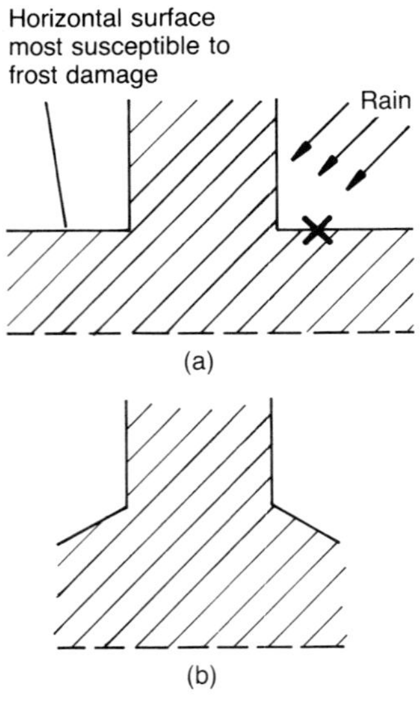

Fig. 12.2. (a) Horizontal surfaces on which water sits should be avoided; (b) such surfaces should be sloped if possible

12.1.1.4. Concrete technology and frost. When selecting the components of the concrete mix, it should first be checked whether the aggregates provided are frost-resistant.

With respect to cement type, the practical measures normally advised to prevent scaling or total degradation are mainly based on experience with ordinary Portland cement. Where severe frost attack has to be taken into account (involving water-saturated concrete or de-icing salts) special precautions may be necessary when using blended cements or blending agents, to prevent scaling. Whether additional precautions may be necessary will depend on the quality of the blending agent, the amount added (into the cement or into the concrete), the W/C ratio and the curing regime. Universally valid figures cannot be given because the quality and quality criteria for blended cements and blending agents differ from country to country, as does the approach used. Some countries have standards both for blended cement and blending agents; other countries have only standards for blended cements. As a rough guide, however, one should be careful when the amount of blending agent exceeds 50% for slag, 15% for fly ash and 8% for silica fume (the percentage being based on the sum of the amount of clinker plus the amount of blending agent) when the concrete will come into contact with de-icing salts. In the case of long cold periods, blended cements may be advantageous in preventing total frost degradation. If these limit values are to be exceeded, the frost resistance of the concrete provided for the particular structure has to be checked by means of relevant tests.

If drying out of the concrete during freezing is guaranteed, a W/C ratio of less than 0·60 and a cement content of greater than 270 kg/m^3 are suitable values to achieve a sufficiently high frost resistance. If, however, water saturation cannot be excluded, the W/C ratio should not exceed 0·55 and the concrete should contain at least 300 kg/m^3 of cement and artificial air pores. These measures are also adequate to limit the risk of capillary suction into the concrete as far as immersed structures and frost attack are concerned.

The air content should be adapted to the severity of attack (e.g. ranging from approximately 3·5% in central Europe to 5·5% in northern Europe) and, in the event of any severe attack (extreme temperatures or frost and de-icing salt attack), be at least 5%. Maximum particle sizes of the aggregate smaller than 32 mm require the air content to be increased, by up to a further 2·5% at 8 mm particle size.

12.1.1.5. Execution and curing in relation to frost. Any frost attack will start at the surface of the concrete. Therefore, the quality of the outer layers of the concrete, and thus curing, will be of major influence on the resistance of the concrete to freezing, i.e. the capacity of the concrete to withstand repeated freezing and thawing.

All curing measures are aimed at preventing premature drying out in order to ensure a high degree of hydration. However, concrete will become resistant to freezing only if a certain degree of drying has been obtained after manufacture, i.e. if part of the excess water resulting from the manufacturing process has been disposed of. For this reason, measures have to be taken to ensure that concrete is not subjected to freezing during the curing procedure and for a sufficient period of time after curing has been completed (see Appendix 1).

The concrete may be assumed to have sufficient resistance against early freezing (i.e. freezing during the curing period) provided it has obtained a minimum compressive strength of 5 MPa at the onset of freezing.

12.1.1.6. Special measures against frost for structures in use. If the concrete quality fails to meet the requirements owing to incorrect planning or execution, or if frost resistance cannot be expected under the prevailing ambient conditions, special measures may become necessary to prevent destruction of the concrete and the structure.

All such measures should aim at avoiding saturation of the concrete. This will be achieved most effectively by preventing capillary suction and the penetration of salts.

For this purpose, the surfaces subjected to wetting may be either impregnated (elimination of the surface energy within the pores) or provided with a coating. However, it should be noted that as a rule impregnations (and to an extent coatings) are not durable and therefore will have to be replaced at regular intervals. In any case, regular inspection and checking of the measures with regard to their efficiency will be necessary.

It is of further importance that all the surfaces exposed to water be treated; otherwise, if certain surface areas remain untreated (e.g. the bottom of foundations), the concrete is likely to become saturated by capillary suction even in the region of surfaces already treated.

12.1.1.7. Erosion. A high percentage of coarse aggregates consisting of wear-resistant rock, held together by a high-strength cement mortar (low W/C ratio) ensuring a good bond of the aggregates, will resist abrasive wear, provided that the surface layer of sealing mortar is thin and the curing has ensured a crack-free surface.

Powdered carborundum or corundum are sometimes used as aggregates in concrete for thin layers on steps, floors or similar structures subject to intensive wear.

A smooth, strong concrete surface with a dense high-strength cement paste ensuring a good bond to the coarse aggregates will resist erosion due to cavitation, provided an adequate curing has rendered the surface crack-free. The design should be such that high streaming velocities at discontinuous profiles should be avoided; this gives an optimal hydraulic design.

12.1.2. Protection against chemical attack

Preventive measures are a function of the degree of aggressivity of the environment (see Tables 9.1 and 9.3), but in all cases a concrete of low permeability, well designed and well made, rarely deteriorates.

Measures for the mix design of concrete, taking into account the type of cement, the W/C ratio and the cement content of the concrete, are given in Tables 12.1 and 12.2. These tables will form valuable supplements to one another until a more coherent system has been developed.

In some cases, additional protection of concrete is necessary. Simple rules can be used to determine the choice of the constituents and the placing of concrete (Table 12.2). A detailed discussion of concrete degradation and proposals for protection are given by Biczok.[26]

In an environment which dissolves calcium products, composite or blended Portland cements (blast-furnace slag cements or pozzolanic cements) are better than Portland cements with a high amount of $3CaO \cdot SiO_2$ (which liberates

Table 12.2. Measures to be taken against chemical attack of concrete by waters and soils containing aggressive agents[15]

Concrete parameter	Exposure class* 5a			Exposure class* 5b	Exposure class* 5c	
	Weak attack			Moderate attack	Strong attack	Very strong attack
Aggressive agents in which sulphates are present						
Type of cement†	OC	OC	SRC	SRC	SRC	SRC
Maximum W/C ratio	0·55	0·50	0·55	0·50	0·45	0·45
Minimum cement content: kg/m³	300	330	300	330	370	370
Additional protection of concrete	Not necessary					Necessary
Aggressive agents in which sulphates are not present						
Type of cement†	OC			OC	OC	OC
Maximum W/C ratio	0·55			0·50	0·45	0·45
Minimum cement content: kg/m³	300			330	370	370
Additional protection of concrete	Not necessary					Necessary

* See Table 9.1.

† OC = ordinary cement; SRC = sulphate-resistant cement (as denominated by the standards of the relevant country). Aggregates should have a maximum size of about 30 mm.

Table 12.3. Guidelines for sulphate resistance of concrete

Degree of attack (see Table 9.3)	Protection mechanism	Protective measure
Weak	Permeability	Maximum 50 mm water penetration (RILEM method)
	W/C	Maximum of 0·6
	Protective coating	
	Type of cement	High sulphate-resistant cement
Moderate	Permeability	Maximum 30 mm water penetration (RILEM method)
	W/C	Maximum of 0·5
	Protective coating	
	Type of cement	High sulphate-resistant cement
Strong	Permeability	Low water penetration
	W/C	Maximum of 0·4
	Protective coating	
	Type of cement	High sulphate-resistant cement
Very strong	—	Not treated

relatively large amounts of calcium ions during hydration). The addition of silica fume seems efficient in this respect.

12.1.2.1. Sulphate attack. Guidelines for measures to be taken are given in Table 12.2. More detailed recommendations are given in Table 12.3.

Due to the equilibrium of the chemical reaction, the formation of ettringite diminishes from a maximum value to zero over the temperature range 0–80°C. As a practical consequence, sulphate corrosion of concrete does not show the usually accepted acceleration in hot climates.

The combined action of sulphate and chloride is dealt with in section 13.3.

12.1.2.2. Alkali reaction. Although progress has been made in evaluating problems concerning this kind of concrete degradation, a comprehensive and satisfactory treatment of the subject is not yet available. However, a safe approach would be, whenever possible, to choose non-reactive aggregates.

Another safe approach seems to be the use of low-alkali cement, as formulated in several national standards, provided that influx of alkali from the exterior (e.g. de-icing salts) is prevented. The American Society of Testing and Materials limits alkali content, expressed as equivalent sodium oxide ($0{\cdot}658K_2O + Na_2O$), to 0·6% by weight. Furthermore, proper allowance should be made for possible influx of alkali from the exterior. Exposure conditions seem to be underestimated in practice. Intermittent drying and wetting may lead to greater expansion; waterproofing may prevent or sufficiently retard expansion.

The use of pozzolanic admixtures, especially silica fume, seems profitable when reducing expansions due to alkali-silica reactions, thanks to their alkali binding property.

Recent opinion has tended to accept Portland blast-furnace cement with a minimum of 65% slag and pozzolanic cements with a minimum of 30% pozzolanic material (either natural or synthetic) as sufficient protection with any kind of aggregate, regardless of the alkali content of the cement.

The addition of some pozzolanas increases the water requirement of the mix, if plasticizers are not used. It should be noted that pozzolanas are not effective in controlling the alkali-carbonate reaction.

Air entraining has been found to be effective in reducing expansion due to alkali-silica reactions.

A low W/C ratio giving rise to a strong concrete with a low permeability, and self-desiccation of the hardened concrete, diminishes the risk of alkali-aggregate reactions.

The reactivity of siliceous aggregate is affected by its particle size and porosity. An appendix on the methods for evaluating potential reactivity is given in ASTM C33.[27] A petrographic description supplies the first indication; an exhaustive procedure is given in ASTM C295.[28] A chemical test, such as ASTM C289,[29] may screen out non-susceptible material, although positive test results may include harmless materials. A mortar bar test, following ASTM C227,[30] is most generally used, although its usefulness is still under discussion. Its main drawback is that it takes at least six months to carry out; some materials may proceed to deleterious expansion even at later ages. It has been suggested that concrete prisms provide a better test for slowly expanding rocks.

The chemical test is useless when estimating alkali-carbonate reactivity. A rock cylinder method, specifically for alkali-carbonate reaction, is given in ASTM C586;[31] this is comparable with the proposed concrete prism test for carbonate rock and slowly expanding silicate rock.

Table 12.4 summarizes the possible tests for alkali-aggregate reactivity.

12.1.3. Protection against biological attack

Biological growth is dealt with in section 11.2. Due to the ability of biological growth to increase the moisture content on the concrete surface thereby give rise to increased deterioration, and due to the risk of mechanical damage caused by the roots entering cracks and voids, the amount of biological growth should be minimized.

The deterioration of concrete due to sulphur from bacteria can be decreased by minimizing the turbulence in sewer pipes, thus reducing the release of

Table 12.4. Guidelines for concrete resistance against alkali-aggregate reactivity

Testing of aggregate	Protection mechanism if test results negative or suspect	Protective measure
General procedure[27]	Limiting of total alkali	Low alkali cement Portland cement $<0{\cdot}6\%$ equivalent Na_2O
Petrographic description[28]	Type of cement	Blended cements Portland blast-furnace cement $>65\%$ slag Pozzolanic cement $>30\%$ pozzolan
Chemical test[29] — fast, but limited applicability; not for carbonates	Sufficient fines in concrete mix	Modified concrete mix fines added mix proportion of (reactive) aggregate
Mortar bar test[30] — preferably on actual mix	Increased proportion of reactive material (at safe distance from pessimum)	
Concrete cylinder test[31] or concrete prism test for carbonates or slowly expanding silicates	Limiting of available water	Low W/C ratio Crack sealing and waterproofing to prevent progression of deterioration

hydrogen sulphide, and by removing the growth of the bacteria on the inside of the sewer pipe. If the circumstances allow, good ventilation of sewers is an efficient way of preventing this process (see chapter 5).

12.2. Protection of reinforcement

When considering the sensitivity of reinforcement to corrosion, the model code[6] distinguishes between two types: sensitive and slightly sensitive. Reinforcement types sensitive to corrosion are

(*a*) steels of all types with diameter ≤4 mm
(*b*) treated steels of any diameter (except quenched and tempered ordinary reinforcement)
(*c*) cold-worked steels subjected to a permanent tension exceeding 400 MPa.

All other types of reinforcement are considered slightly sensitive to corrosion.

The characterization of exposure classes related to environmental conditions affecting the reinforcement (Table 9.2) leads to the need for a corresponding set of requirements for minimum concrete cover. It is suggested that Table 12.5, based on Table 7.2, can be used with the definitions given in Table 9.2. This table also covers prestressed reinforcement, including pre-tensioned tendons with direct bonds and post-tensioned tendons in grouted ducts.

The spacers should be designed so that the nominal cover c_{nom} is 5 mm greater than the minimum cover c_{min}. To ensure fire resistance, higher values of concrete cover may be required.

Although use of concrete with higher strengths or lower W/C ratios than the required values for the different exposure classes would need slightly lower covers compared with those given in Table 12.5, in order to minimize errors this is not recommended.

12.2.1. Planning

12.2.1.1. Structural detailing. A great number of cases of damage are caused by weak points of the structure. During planning, the following points should be considered (see also chapter 10).

(*a*) Concrete surfaces should be as smooth as possible. Near edges, aggressive agents will act from two sides. Tolerances in concrete cover may be found on both sides, and the danger of spalling is increased.
(*b*) Saturation with water should be avoided. Adequate drainage should be provided, there should be no horizontal concrete surfaces and special measures of protection should be taken if necessary.
(*c*) The replaceability of structural elements after severe attack by aggressive agents should be a consideration.

12.2.1.2. Limitation of crack widths for ordinary reinforcement. In the region of cracks, depassivation of ordinary reinforcing steel must be avoided. The absolute value of crack widths w in the normal range ($w = 0{\cdot}4$ mm) is of minor importance compared with the quality of concrete cover (thickness and impermeability). Crack-width limitation using Table 12.6 is normally sufficient.

Table 12.5. Minimum cover

Exposure class (see Table 9.2)	C_{min}: mm	
	Ordinary reinforcement	Prestressed reinforcement
1	15	25
2	30	35
3,4	40	50
5	—*	—*

* Depends on the type of environment encountered.

Table 12.6. Reinforcement detailing rules for ordinary reinforced concrete structures to limit crack widths[32]

Steel stress under quasi-permanent loading: MPa	160	200	240	280	320	360	400	450
Maximum bar diameter: ϕ_s* mm†	—	25	20	16	12	10	8	6
Maximum bar spacing: mm								
Pure flexure	250	200	150	100	60			
Pure tension	125	100	75	50	—			

† The maximum bar diameter ϕ_s may be adjusted for load-induced cracking to $\phi_s = \phi_s^* h/10(h-d)$ and for restraint-induced cracking to $\phi_s = \phi_s^* f_{ctm}/2{\cdot}9$, where ϕ_s^* is taken from Table 12.6, h is the overall depth of the section, d is the effective depth and f_{ctm} is the concrete tensile strength.

Of major importance is the provision of a minimum amount of reinforcement in the case of restrained deformations, to avoid extremely wide cracks.

In the case of very severe environmental conditions (e.g. severe chloride attack on horizontal concrete surfaces), high corrosion rates may occur in the region of cracks. Again, limitation of crack width is not sufficient to avoid the attack on the reinforcement. In such cases (e.g. car park decks) special protective measures must be taken (e.g. sealing the concrete surface or the use of epoxy-coated reinforcement).

12.2.1.3. Limitation of crack widths for prestressing steel. The design principle here is completely different from that for ordinary reinforcement. Due to the danger of brittle failures, depassivation of the prestressing steel surface must be avoided during the entire lifetime. For this reason, the durability of prestressed members may be more critically affected by cracking.

Cracks crossing the prestressing steel in outdoor conditions can only be allowed when the members are post-tensioned (additional protection is provided by the duct and the grout), there is no chloride attack and the crack width at the concrete surface $w_k \leq 0{\cdot}2$ mm. In all other cases decompression must be asked for, with the additional requirement that, under frequent combinations of loads, all parts of the tendons or ducts lie at least 25 mm within concrete in compression (Table 12.7).

12.2.1.4. Detailing of reinforcement. When detailing the reinforcement, the practicality of appropriate concreting and compaction should be allowed for. Especially in the case of crossing layers of reinforcing bars, gaps for the insertion of a vibrator should be provided. When plotting the bar spacings, it should be taken into account that the bars, including the ribs, have greater diameters than the nominal values and that, especially in the case of bent bars, tolerances are necessary.

12.2.1.5. Concrete cover and reinforcement spacings. Minimum values for the concrete cover depending on environmental conditions have been given earlier in this chapter. The nominal values of concrete covers should include allowances for a tolerance value. This ensures that the specified minimum

Table 12.7. Crack-width limitation criteria for prestressed members

Exposure class (see Table 9.2)	Design crack width w_k under the frequent load combination: mm	
	Post-tensioned	Pre-tensioned
1	0·2	0·2
2	0·2	Decompression*
3	Decompression* or coating of the tendons and $w_k = 0{\cdot}2$	
4		

* All parts of the tendons or ducts must lie at least 25 mm within the concrete in compression under the frequent combination of loads.

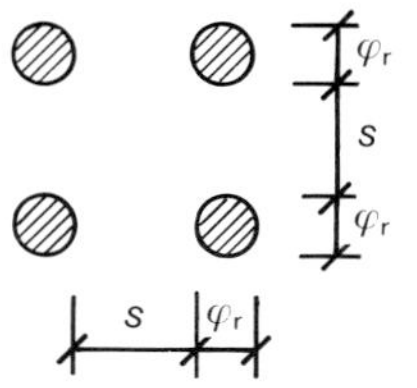

Fig. 12.3. Bar spacings. $\phi_r = 1{\cdot}2 \times$ *nominal* ϕ; $\phi_r \geq \phi$ *or* ϕ_n *(for bundles)*; s $\geq$ *20 mm (the nominal value of* s *should not fall below 30 mm if possible)*; s $\geq 1{\cdot}5 \times$ *maximum aggregate size*

values will be observed everywhere in the structure. The tolerance value depends on quality control during execution and the type of production, e.g. pre-casting. With adequate quality control and curing it should be 0·5 cm; without quality control it should be increased to 1·0 cm, and if the curing is inadequate, 2·0 cm should be allowed.

Because of the increase in effective bar diameter caused by the ribs, and the tolerances associated with bending, the specified minimum values for the bar spacing should be increased correspondingly (Fig. 12.3).

12.2.1.6. Concrete composition. The W/C ratio should be $\leq$0·5 (0·4 if severe chloride attack is expected), and the cement content should be ≥ 300 kg/m^3, unless suitable tests show that for special conditions other limit values may be used. The W/C ratio and the workability, which should be semi-fluid or fluid,[6] are of major importance.

In the case of severe chloride attack, a high-strength concrete with blended cements (slag, natural pozzolanas, fly ash, silica fume) (characteristic concrete strength $f_{ck} \geq 35$ MPa) with a W/C ratio $\leq$ 0·4 and an increased cement content provides increased protection of reinforcement. This increased protection, however, can only be achieved if special curing measures are guaranteed. Tests to prove the mix design are recommended.

12.2.1.7. Critical chloride content. The critical chloride content, indicating incipient danger of corrosion, depends on various parameters. There is therefore no single generally valid value of critical chloride content. The situation is shown in Fig. 12.4.

If concrete is not carbonated, 0·05% Cl^- related to the weight of concrete, or say 0·4% Cl^- related to the cement weight, is a good criterion for incipient danger of corrosion. However, as Fig. 12.4 shows, the critical value can be much higher or lower, depending on the environmental influences.

As prestressing steels are more sensitive to corrosion, a lower limit of about 0·025% Cl^- by weight of concrete, or say 0·2% Cl^- related to the cement weight, is recommended for prestressed structures.

Chlorides must not be added deliberately to the concrete mix, regardless of whether or not an agreed maximum chloride content would be exceeded.

12.2.1.8. Severe chloride attack. In the case of severe chloride attack (e.g. in a splash-water zone, or in hot and wet climates), the measures recommended above may not be sufficient to ensure appropriate durability. The cheapest measure is to decrease the W/C ratio even more and to increase the concrete cover. The use of high-strength concrete with blended cements

Fig. 12.4. Variation of critical chloride content with environment

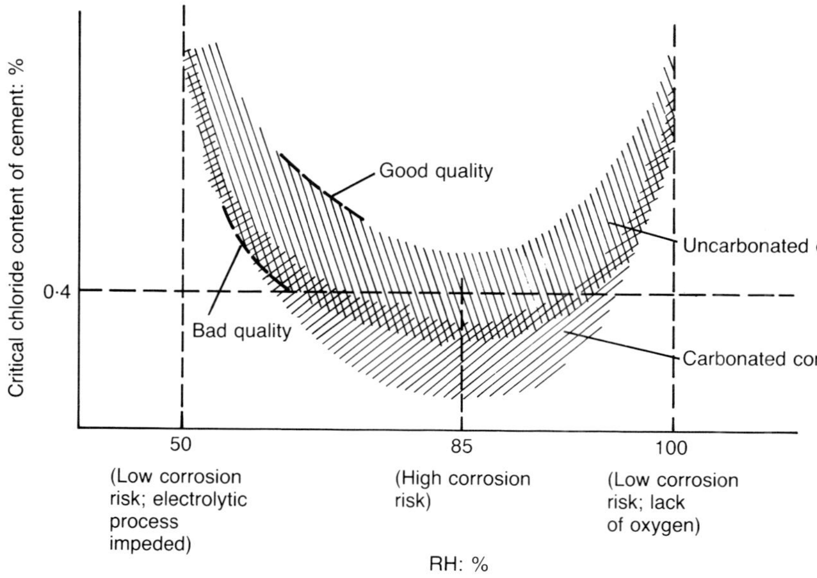

Table 12.8. Spacing of spacers: values for s_0 (see Fig. 12.5)

ϕ_s: mm	s_0: mm
≤8	400
8–14	500
>14	700

Table 12.9 (right). Spacing of spacers: values for s_2 (see Fig. 12.5)

ϕ_s (longitudinal reinforcement): mm	s_2: mm
≤10	500
12–20	1000
>20	1250

Table 12.10 (right). Spacing of spacers: values for s_1 (see Fig. 12.5)

Spacer type	s_1 (horizontal position, e.g. slabs)	s_1 (vertical position, e.g. walls)
Single spacer	s_0	$1{\cdot}5\ s_0$
Spacer system	$1{\cdot}5\ s_0$	$2{\cdot}0\ s_0$

may be advisable (see section 12.2.1.6). If the chloride attack is combined with a severe freeze-thaw attack, however, special precautions should be taken when applying composite or blended cements (see section 12.1.1.4).

If these precautions appear to be insufficient, it may be most suitable to protect the reinforcement directly, e.g. by using epoxy-coated reinforcement (in the design of new structures) or by using cathodic protection (for protecting existing structures or as a preparatory measure for new structures).

Galvanizing and inhibitors have not proved to be very effective in improving durability in the case of severe attack. Epoxy coatings and cathodic protection seem to be more effective.

12.2.2. Execution

12.2.2.1. Concrete cover — spacing of spacers. Even if tolerance values are added to the nominal values of concrete cover, the specified minimum values can only be met if stable spacers are provided with sufficiently small spacings. The minimum values given in Tables 12.8–12.10 must be observed; the quantities are defined in Fig. 12.5.

12.2.2.2. Curing of concrete. The curing measures must ensure that early drying-out of the concrete surface is not allowed to take place. Once the concrete has dried out, any subsequent curing measures will be useless. The following may be used.

(*a*) formwork
(*b*) covering of surfaces without formwork (curing overlays)
(*c*) temporary coatings or impregnations.

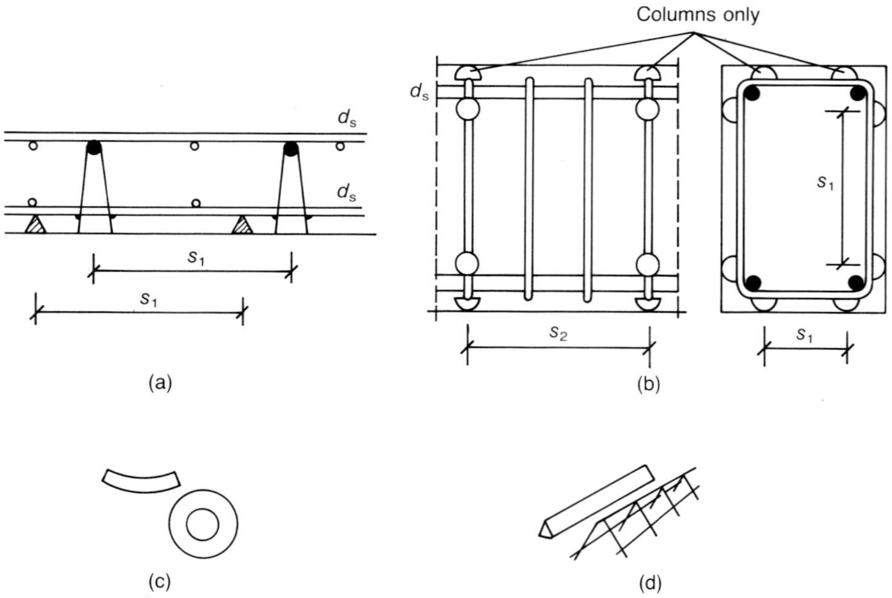

Fig. 12.5. Spacers, showing the quantities specified in Table 12.8: (a) slabs and walls; (b) beams and columns; (c) a single spacer; (d) spacer systems

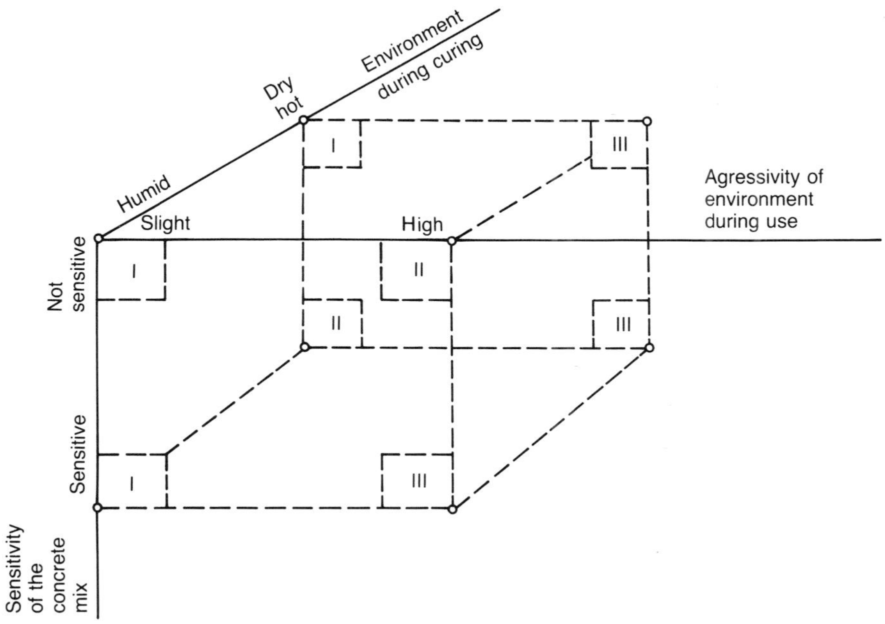

Fig. 12.6. Recommended curing times: I. 1–3 days; II. 5–7 days; III. 10–14 days

The required curing time depends on several parameters. The main ones are

(*a*) the aggressivity of the environment during the service life
(*b*) the environment during curing
(*c*) the curing sensitivity of the concrete mix (type and amount of cement, and W/C ratio).

Figure 12.6 gives advice on the required curing time depending on these parameters (see also Appendix 1). However, to arrive at more robust concrete mixes, the sensitivity of the mix to curing should be decreased by reducing the W/C ratio. In this way Fig. 12.6 can be reduced close to a two-dimensional diagram.

12.2.2.3. Quality control. The measures described must be checked in an appropriate manner by the responsible engineer on the site. The following in particular should be checked.

(*a*) bar spacings, gaps for inserting of vibrator
(*b*) concrete cover, suitability and spacing of spacers
(*c*) mix design and quality assurance system for concrete quality
(*d*) measures for curing.

13. Measures to cope with typical environments

In practice a multitude of coinciding aggressive factors of varying intensity are present, thus seriously complicating the task of making the right decisions about materials, techniques and procedures influencing the service life of structures. A first step towards handling the complexity of actual environments is given in this chapter.

13.1. Indoor environments

Indoor environments are as variable as the uses to which buildings may be put. Generally, the aggressivity of the environment is low in structures used for general occupancy, offices, schools, etc. As far as corrosion is concerned, it is generally reasonable to consider an average relative humidity of around 50%. It should be noted, however, that the temperature is liable to be around 20°C. This gives a minimum cover of about 15 mm. Care should be taken to identify any areas where condensation might occur on concrete surfaces and either increase the cover or protect the surface. Condensation will increase the effective relative humidity and thus increase the corrosion risks.

It is hard to say anything useful about indoor environments, as the range of possibilities is so large. A basic procedure might be

(*a*) to establish clearly the nature of the usage expected for the structure
(*b*) to assess what this means in terms of effective average humidity for the various parts of the structure
(*c*) to assess the expected average temperature regime
(*d*) to decide whether or not there are deleterious substances (e.g. chlorides and acid gas) which may frequently be in contact with the concrete.

This should permit identification and quantification of the risks. Appropriate action can then be taken, using the information given in previous chapters.

13.2. Outdoor environments

As with indoor environments, it must be emphasized that there is no single outdoor environment: in the world there are enormous variations in humidity, rainfall and temperature. Even at the most local level, there are substantial differences in environment between the sheltered and more exposed areas of a single structure. It will be seen that while definitions of environment such as those quoted in chapter 9 may be barely satisfactory when applied to a single country, they cannot be adequate on an international basis.

A further feature of outdoor environments should also be noted: the variations in the microclimate around an external member will be great. The microclimate will vary with the weather, with the seasons, and with the time of day or night. Internal environments tend to be much more uniform.

Little more of direct help can be said, except that the most appropriate approach would seem to be that set out in section 13.1 above, and to follow some basic rules about structural form which are outlined below. These follow from the principle that saturated concrete is most at risk from frost attack, alkali-aggregate reaction attack and any other form of chemical attack, if the appropriate chemicals are present.

Corrosion is not a great risk if the concrete can remain permanently fully saturated, but in many environments, particularly those where chlorides are present, increases in effective relative humidity due to wetting and drying lead to more corrosive conditions. It is therefore important to avoid structural arrangements which do not allow easy drainage of rainwater from all structural surfaces.

It is particularly important to avoid arrangements which may allow water contaminated with chlorides to drain over structural concrete surfaces (an example of this, which has been the cause of much trouble, is where water containing de-icing salts has been allowed to drain through faulty joints in bridge decks and over the supporting structure).

Therefore it is necessary

(*a*) to avoid horizontal surfaces on which water may stand
(*b*) to ensure that drainage is provided which will prevent water from being channelled over the surface of structural concrete, particularly if this water might contain chlorides
(*c*) to ensure that the drainage arrangements can be easily maintained and that they are cleaned frequently
(*d*) to allow for the possibility of joints leaking.

For further details see chapter 10.

13.3. Concrete in contact with soils

In the common case of a combined attack by sulphates and chlorides — and high temperature and sometimes high relative humidity, as in the Middle East — the choice of tricalcium alumina content becomes difficult. As blended cements have a low permeability for chlorides, the best compromise seems to be to specify a dense, homogeneous concrete made from a low tricalcium alumina cement plus slag or pozzolanic materials, or equivalent blended cements, when a combined sulphate and chloride attack is expected.

Soil contaminated with mineral oil may be aggressive to concrete if the oil contains acidic components (i.e. phenols or organic acids).

13.4. Concrete in a marine environment

13.4.1. Nature of the environment

13.4.1.1. Constituents of sea water. Sea water contains many dissolved salts, some of which affect the durability of concrete. The salts which are present in significant quantities in most seas are sodium chloride (NaCl), magnesium chloride ($MgCl_2$), magnesium sulphate ($MgSO_4$), calcium sulphate ($CaSO_4$), potassium chloride (KCl) and potassium sulphate (K_2SO_4). Concentrations vary from sea to sea, although the total salt content is commonly about 35 g/l. An exception to this is the Baltic, which contains only about 1/5 of this amount of dissolved salts. Figure 13.1 gives typical

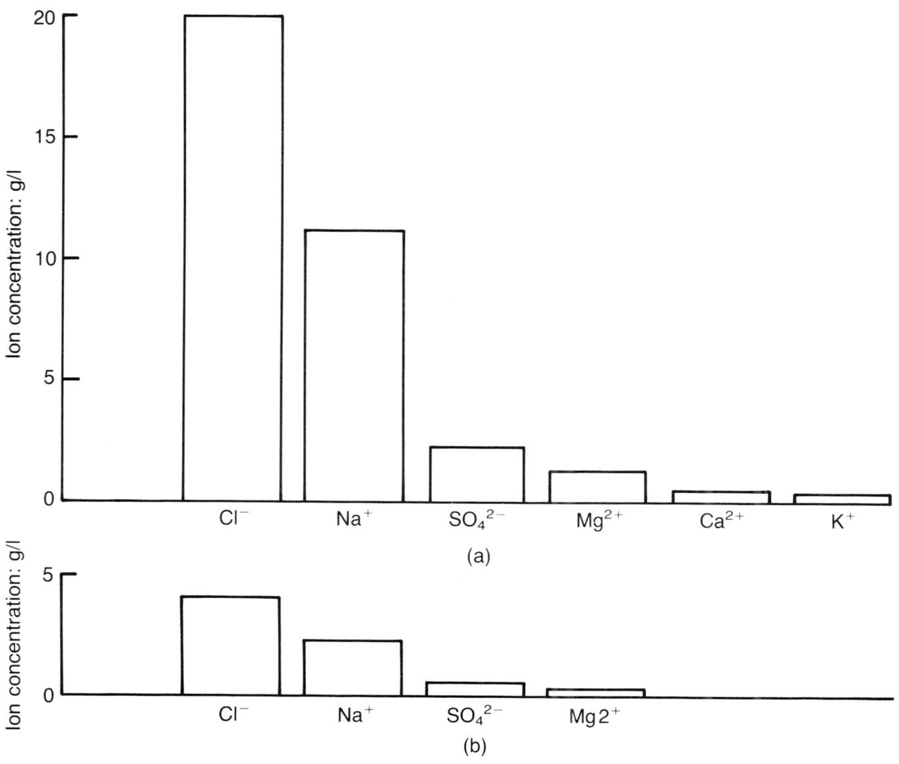

Fig. 13.1. Typical ionic concentrations of the commoner salts: (a) in the Atlantic Ocean; (b) in the Baltic Sea

Fig. 13.2. Types of marine exposure

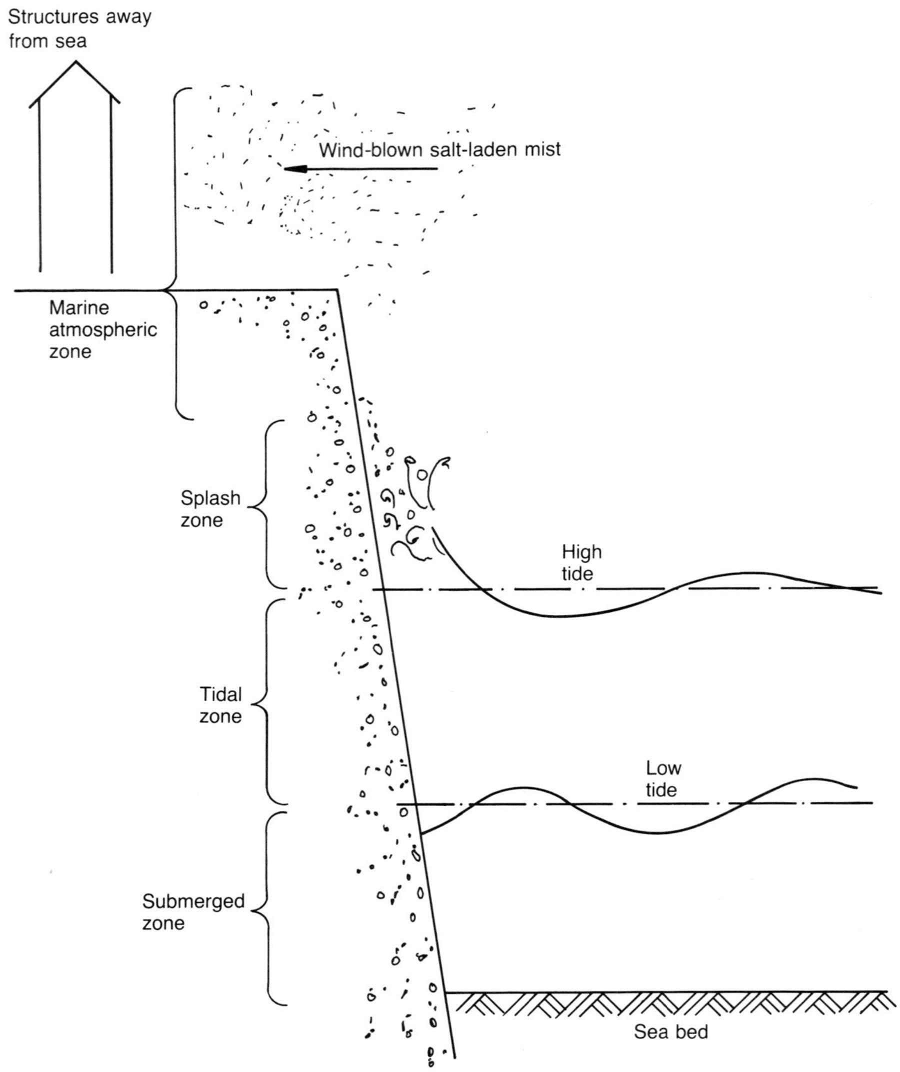

Fig. 13.3. Deterioration of concrete structures in sea water

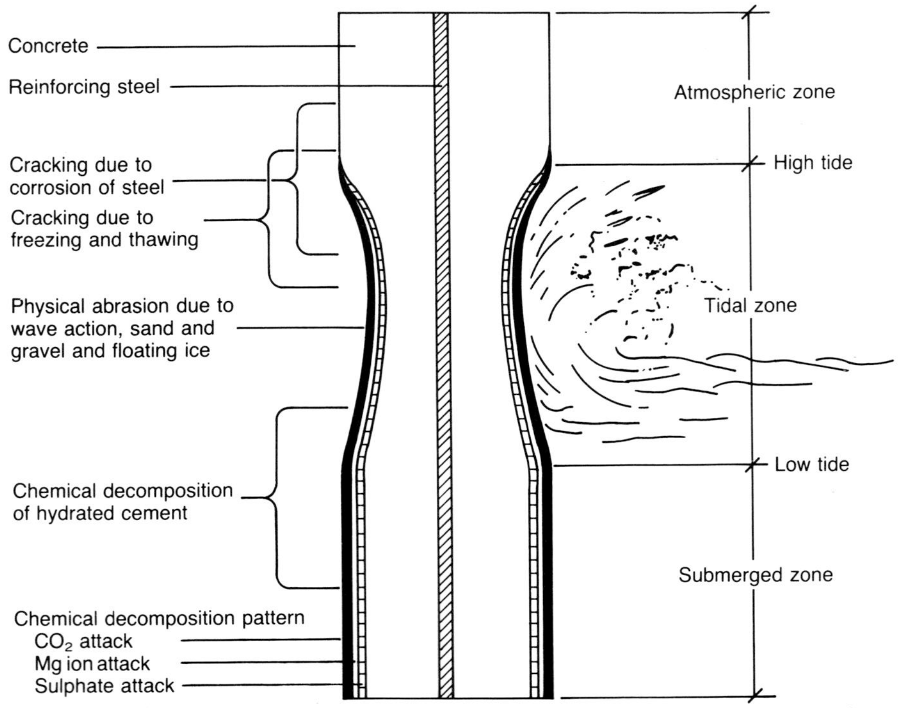

ionic concentrations of the constituents of the commoner salts for the Atlantic and the Baltic.

It should be noted that sea water also contains dissolved oxygen and carbon dioxide. The concentrations of these gases can be highly variable, depending on local conditions.

13.4.1.2. Basic exposure zones. There are several different types of marine exposure, each with its own particular characteristics and hazards (Fig. 13.2). They may be summarized as

(*a*) the marine atmospheric zone in which concrete is never directly in contact with the sea although it will receive salt from blown spray and salt-laden mist; the chloride levels will decrease with increasing distance from the sea but, depending on the nature of the coast and prevailing winds, salt may be blown many kilometres inland
(*b*) the splash zone, which lies above high tide but is still subject to direct wetting by sea water from waves and spray
(*c*) the tidal zone, which lies between high and low tides; concrete will be submerged for periods each day
(*d*) the submerged zone, which is below low tide and in which concrete is continuously submerged
(*e*) the seabed zone.

It should be noted that definite boundaries do not generally exist between these environments; one zone tends to merge into the next.

13.4.2. Possible causes of deterioration in the various zones

The forms of deterioration (Fig. 13.3) which are most prevalent may be summarized as

(*a*) marine atmospheric zone
  (i) corrosion of reinforcement activated by chloride
  (ii) frost damage
(*b*) splash zone
  (i) corrosion of reinforcement activated by chloride
  (ii) abrasion due to wave action
  (iii) frost damage
(*c*) tidal zone
  (i) abrasion due to wave action, floating ice and other objects, ship collision and so on
  (ii) corrosion of reinforcement activated by chlorides
  (iii) frost damage
  (iv) biological fouling
  (v) chemical attack on concrete
(*d*) submerged zone and seabed
  (i) chemical attack on concrete
  (ii) biological fouling and attack by organisms.

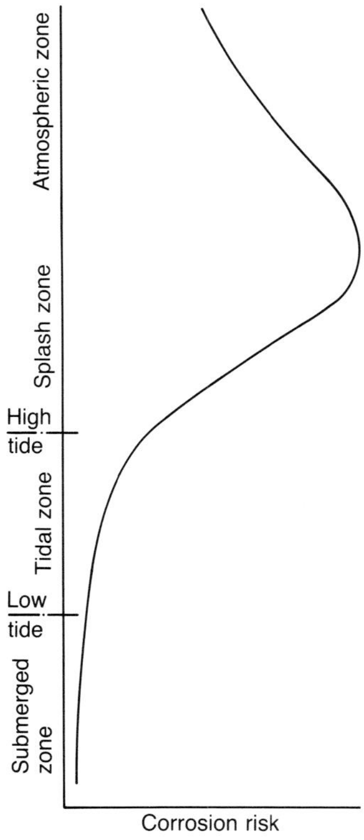

Fig. 13.4. Variation of corrosion risk across the marine exposure zones

A number of these mechanisms are now treated in more detail.

13.4.2.1. Reinforcement corrosion. Experience suggests that the greatest risk of corrosion of the reinforcement occurring is in the splash and atmospheric zones. The risk decreases rapidly with the distance below high tide and is very low in the submerged zone (Fig. 13.4). However, in the case of very porous concrete with low cement content, heavy chloride corrosion in the submerged zone has been observed (macrocell corrosion).

The usually very low risk of corrosion damage in the submerged zone is due to the low concentration of oxygen in the water and the slow rate at which it can diffuse through the water-saturated concrete to the steel.

There is a much greater abundance of oxygen in the tidal zone, but corrosion is still limited by the slow rate of diffusion through saturated concrete.

13.4.2.2. Chemical attack. The various modes of chemical attack which are known to occur are summarized in Fig. 13.5. Chemical deterioration of the concrete due to these mechanisms is only likely in the lower part of the tidal zone and the submerged and seabed zones.

Chemical attack by sea water is not commonly a serious problem in concrete of reasonable quality. One reason for this is that the expansive behaviour associated with ettringite formation is inhibited by the presence of chlorides. In order to reduce the risk of sulphate attack, some authorities have specified upper limits to the tricalcium alumina content of cement used for marine works. Such advice should, however, be viewed with caution since cements without tricalcium alumina (e.g. sulphate-resisting cements) are less able to protect reinforcement from the ingress of chlorides.

13.4.2.3. Fouling and biological attack. Marine growth (fouling) can be

Fig. 13.5. Chemical processes involved in the deterioration of concrete by sea water[16]

1. Action of CO_2

(*a*) $Ca(OH)_2 + CO_2 + H_2O \rightarrow CaCO_3 + 2H_2O$

Precipitate

Aragonite — Calcite

[Coating]

2. Action of sulphate ($MgSO_4$)

(*b*) Substitution of Mg^{2+} by Ca^{2+}

$MgSO_4 + Ca(OH)_2 \rightarrow CaSO_4 + Mg(OH)_2$

Soluble (1·2 g/l) — Solid secondary gypsum — Precipitate

[Leaching] — [Expansion] — [Coating]

(*c*) Action of secondary gypsum

$CaSO_4 + \text{alumina} + 32H_2O \rightarrow (\text{alumina})\cdot 3CaSO_4\cdot 32H_2O$

Ettringite

[Expansion]

3. Action of chloride ($MgCl_2$)

(*d*) Substitution of Mg^{2+} by Ca^{2+}

$MgCl_2 + Ca(OH)_2 \rightarrow CaCl_2 + Mg(OH)_2$

Soluble — Precipitate

[Leaching] — [Coating]

(*e*) Action of $CaCl_2$

$CaCl_2 + \text{alumina} + 10H_2O \quad (\text{alumina})\cdot CaCl_2\cdot 10H_2O$

Chloroaluminate

[Expansion]

SO_3

$(\text{alumina})\cdot 3CaSO_4\cdot 32H_2O$

Ettringite

[Expansion]

$CO_2 + SiO_2$

$CaCO_3\cdot CaSO_4\cdot CaSiO_3\cdot 15H_2O$

Thaumasite

[Expansion]

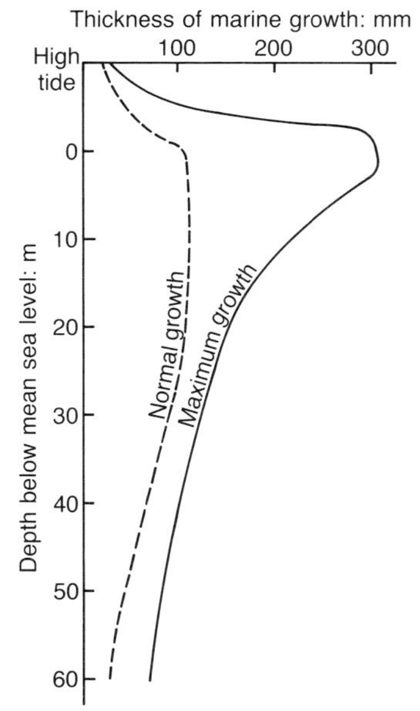

Fig. 13.6. Variation of thickness of marine growth with depth

of significance on some types of structure. The effect is mainly physical: it increases drag and can increase the forces induced in some types of structure by wave action by up to 100%. The inertia of members covered in marine growth can also be considerably increased.

Figure 13.6 indicates the thickness of marine growth which can occur. Maximum growth occurs where large quantities of nutrient are available, for example near a sewage outfall.

Actual damage to concrete by marine growth organisms is not commonly a problem. It has been reported that seaweed can increase the rate of degradation of concrete. This is likely to be due to the action of organic acids and sulphate produced by decomposing vegetation. It has also been reported that in the tropics some types of mollusc can eat into concrete at a rate of 1 cm per year. Algae in the submerged zone can improve durability by sealing the concrete surface.

13.4.2.4. Other forms of degradation. The risk of damage due to frost attack, abrasion and so on will depend on the particular nature and location of the structure considered.

13.4.3. Practical measures

13.4.3.1. Cement. Where Portland cement is used, consideration should be given to limiting the tricalcium aluminate content of the cement. A maximum of 10% and a minimum of 5% is suggested by some authorities.

Portland sulphate-resisting cement is less able to protect the steel from corrosion than other cements.

Blast-furnace slag cement has a high resistance to both chloride and sulphate attack.

13.4.3.2. Mix proportions. With respect to mix proportions (see section 12.1), the W/C ratio should be kept as low as possible ($<0{\cdot}5$) and the workability should be ensured (e.g. using plasticizers).

13.4.3.3. Cover. In general, the values given in section 12.2 should be applied (exposure class 4). The minimum cover should be increased where abrasion may occur. Less cover may be used in the submerged areas.

14. Appraisal of concrete structures

The investigation and assessment of concrete structures is a task combining all technical and non-technical elements of structural design, construction and materials technology, including aspects of durability, reliability and performance. The level of detail required may vary, from a simple judgement of structural and functional adequacy based on a superficial visual inspection during (regular) inspection rounds, to a profound investigation and evaluation procedure combining special techniques of inspection and testing on both macro and micro levels.

Rational decision models applying modern probabilistic methods including safety policy and economics are indispensable elements of an in-depth appraisal. The investigation may include

(*a*) visual inspection
(*b*) check of original design: drawings and calculations
(*c*) check of execution data available: technical and non-technical; a quality statement record and inspection records (see section 10.1) would greatly facilitate this task

Fig. 14.1. Check-list for investigation of deteriorated concrete[2]

1. Concrete under inspection

A concrete structure
Sample of concrete from a structure (sound and deteriorated)
Laboratory specimen stored on site
Laboratory specimen stored in laboratory
Sampling procedure
Sample storage and treatment

2. Initial data on concrete

Concrete structure (design, dimensions, loading history)
Concrete specifications
Concrete mix design
Tests on materials used
Quality control of fresh concrete
Quality of concrete in place
Duration and type of curing conditions
Age at time of attack

3. Influence from the environment

Temperature
Humidity
Pressure
Permeability of the surrounding media
Sea water
Other aggressive substances
Type of contact
Concentration of aggressive substances
Frequency and duration of exposure
Special environmental influences (stray currents)

4. Visual signs of deterioration

Erosion
Spalling
Exfoliation
Dusting
Crumbling
Softening
Staining
Pop-outs
Cracking
Liquid gel exudation
Crystallization
Corrosion of reinforcement
Misalignment
Others

5. Laboratory examination and tests

Visual examination
Chemical analysis
Thermal analysis
Infra-red spectrometry
X-ray diffraction analysis
Microscopy (optical or electronic)
Mechanical tests
Sonic tests
Dimensional change
Weight change
Capillary absorption
Permeability
Porosimetry
Others

6. Conclusions

Design of concrete structure not appropriate
Concrete specifications not appropriate
Concrete specifications not fulfilled
Control of concrete technology inadequate
Unsatisfactory quality of components

7. Recommendations

Safety precautions
Demolish
Repair
Prevention of recurrence

(*d*) in situ testing: non-destructive, destructive, sampling
(*e*) testing in the laboratory: mechanical, physical, chemical and morphological
(*f*) recalculation.

Decisions regarding safety precautions, repairs, strengthening, upgrading, demolition and prevention of recurrence must be made based on these elements.

CEB General Task Group 19 (Diagnosis and Assessment) is preparing a report on the appraisal of concrete structures[33] and General Task Group 21 is preparing a report on the redesign of repaired or strengthened concrete structures.[34]

A check-list has been established by RILEM to facilitate communication about deteriorated concrete between the building industry and organizations working on the durability of concrete (Fig. 14.1). This check-list may be used in case studies, in long-term studies and in descriptions of concrete failures in practice.

A guide to the investigation of structural failures covering all types of material and construction has been prepared by the American Society of Civil Engineers.[35]

Appendix 1. Curing of concrete structures

Appropriate casting places a number of requirements on the planning and implementation of the work; these apply irrespective of the time of year. This Appendix describes how the hardening process of the concrete can be controlled so that the following requirements are met in a sound manner, with regard to both financial and energy resources.

A1.1. Requirements to be met

Control of the hardening process will usually comprise the following.

A1.1.1. Ensuring an appropriate hardening process
The casting process must be planned so that the required stripping strength is achieved at the required time, taking into consideration the technological, time-related and financial conditions involved in the work.

A1.1.2. Ensuring against damage through early freezing
Experience shows that a hardening concrete will suffer permanent damage if the first freezing takes place before the concrete has matured for 15–20 h. The method chosen must ensure that the concrete will not freeze until the required degree of hardening has been achieved.

A1.1.3. Ensuring against damage resulting from thermal stress
Thermal differences in hardening concrete cross-sections will cause differential movements due to the thermal expansion of the concrete; under unfavourable circumstances this may lead to cracks in the concrete. It must be ensured that these thermal differences are controlled to such an extent that the initial quality of the concrete is not reduced due to cracks.

A1.1.4. Ensuring against damage resulting from drying
The curing of the concrete is part of the hardening process. On the one hand the curing must ensure that too large temperature differences do not occur, and on the other it is meant to prevent premature drying.

A1.2. Basis for planning and control of the hardening process

To understand the principles that form the basis of the planning and control of the hardening process of concrete, attention must be drawn to three items.

(*a*) The rate of hardening is to a large extent determined by the temperature of the concrete. If the temperature of the concrete is raised, the hardening is accelerated. If the temperature of the concrete is lowered, the rate of hardening will decrease. At 35°C the hardening will be approximately twice as fast as at 20°C. At 10°C the rate will be about half of the rate at 20°C.

(*b*) During the hardening of the concrete, heat is developed. In the case of complete hydration of 1 kg Portland cement, about 400–500 kJ will be developed. In a typical concrete mix this will lead to a rise in temperature of about 60–80°C if the concrete is left to harden with no heat loss to its surroundings.

(*c*) The temperature development within a hardening concrete structure is determined by the balance between the development of heat within the concrete and the exchange of heat to the surrounding air. In thick-walled structures or in highly insulated structures the temperature will consequently become high; the heat thus generated does not pass easily to the surrounding air. The situation is quite the reverse in the case of flimsy, uninsulated structures; the generated heat passes easily to the surrounding air, and consequently the increase in temperature is small.

From this it can be seen that the most important factor in the control of the hardening process is the control of the heat generated by the hardening. In practice there are a number of possible ways to control the balance between this generation of heat and the heat exchange between the hardening concrete structure and its surroundings.

A1.3. Control parameters

The balance between the generation of heat and its dissipation in the circumstances is influenced by a number of factors. Generally structural dimensions and weather are given parameters which can either not at all or only to a small extent be influenced by the contractor. Concrete type, cement type and cement content are parameters which can to a certain extent be modified by the contractor for control purposes. Casting temperature, mould type, degree of insulation and stripping time are the actual control parameters in connection with the planning and implementation of casting work.

Which methods are the most appropriate will depend on conditions which may vary from site to site. The choice of method will frequently have financial implications, and consequently time schedules, delays, employment, depreciation of machinery and so on will often determine the method chosen.

A1.4. Elements of curing control

There is a distinction between active and passive control of the hardening process. Various methods of heat curing in the case of industrial production of concrete elements are examples of active hardening control aimed at increasing production. The concrete temperature — and consequently the hardening process — is supposed to be controlled through the addition of heat through steam, infra-red radiation and so on.

As opposed to this, concreting in situ is usually based on passive control of the hardening of the concrete. It attempts to administer the heat developed from the hardening process through the selection of appropriate measures such as casting temperature and mould insulation.

The steep increase in energy prices during the past decade has increased the interest in passive hardening control, during which the generated heat is utilized in the process. In the field of precasting of concrete elements, major improvements have taken place; there are now high-output plants where the heat curing takes place entirely without use of external energy.

Passive hardening control must take as its basis a quantification of the basic factors connected with the hardening of the concrete, the formation of structure, the development of properties, and the heat balance. Thus

(*a*) the influence of temperature on the speed of the hardening process is determined using the appropriate temperature functions
(*b*) the heat development properties of the concrete are measured and described using the appropriate mathematical models
(*c*) the complex heat balance for hardening concrete cross-sections with an internal, non-linear source of heat which depends on location and temperature is handled using numerical calculation routines
(*d*) criteria to achieve resistance against freezing during the early hardening phase are set up by means of the appropriate mathematical models covering the phase composition of the hardening concrete
(*e*) criteria for the duration of the curing of the concrete to prevent drying are also set up on the basis of the phase composition model
(*f*) based on the knowledge of the temperature distribution within the hardening concrete — including the rapid cooling after stripping — operational rules are set up to eliminate the formation of cracks.

A1.5. Dependency on temperature of the hardening process

The rate of the chemical reactions between cement and water will accelerate with increasing temperature. A comparison of the rate at 20°C with the rate at θ°C will give the following approximate proportion

$$E(\theta) = \frac{\text{rate at } \theta\,^\circ\text{C}}{\text{rate at } 20\,^\circ\text{C}} = \exp\left[\frac{E}{R}\left(\frac{1}{293} - \frac{1}{273+\theta}\right)\right] \tag{1}$$

where E is the characteristic activation energy and R is the gas constant 8·314 J/mol °C. For Portland cements, E = 33 500 J/mol for $\theta \geq 20$°C; E = 33 500 + 1470(20 − θ) J/mol for $\theta < 20$°C.

In Fig. 10.21 the temperature function $H(\theta)$ is depicted for temperatures from −10°C to 90°C. $H(\theta)$ shows the temperature dependency of the hardening process in relation to the rate at 20°C. The reference temperature has been chosen to be 20°C because this is the usual reference temperature used in codes of practice and standards.

Through application of the temperature function $H(\theta)$ it is possible to compare hardening processes at a temperature θ with an already known hardening process examined at the reference temperature. This comparison is made by calculating the maturity M of the concrete, which is the equivalent age at 20°C. The maturity at time t is determined by

$$M = \int_0^t H(\theta)\,\mathrm{d}\tau \tag{2}$$

In the case of numerical calculations, the corresponding summation expression is used

$$M = \sum_{i=1}^{n} H(\bar{\theta}_i)\,\Delta t_i \tag{3}$$

The given temperature sequence is divided into n time intervals Δt_i. For each interval, the mean temperature $\bar{\theta}_i$ is determined and the corresponding value of the temperature function $H(\bar{\theta}_i)$ is determined. The addition to the maturity of the concrete M_i in the interval under consideration is determined by $H(\bar{\theta}_i)\,\Delta t_i$. The maturity thus achieved is finally determined by a summation of the calculated maturity additions.

A1.6. Heat development of concrete

The mathematical modelling of the heat-developing properties of concrete under varying temperature conditions is central to the description of the heat balance of hardening concrete structures. Since the heat development is influenced by chemical and mineral additives and W/C ratio, for example, the key data for the mathematical description should as far as possible be determined through measurements made on the concrete in question.

The most appropriate method of determining the heat development properties of a concrete is adiabatic calorimetry. In this method the temperature increase in a concrete test specimen hardening without heat exchange with the surroundings is measured. Once the composition and heat capacity of the concrete are known, the measured temperature increase can be converted into specific heat development in units of kJ/kg cement. Figure A.1 is an example of a suitable form for recording data from adiabatic calorimetry measurements.

When the temperature function $H(\theta)$ is used, the measured heat development can be depicted as a function of the maturity development M. The resulting curve will serve as reference for hardening calculations in the form of temperature versus maturity.[36,37]

If the heat development were to be depicted as a function of the maturity M, an S-shaped curve would appear in the single logarithmic depiction of Fig. A.2. To a good approximation a mathematical representation of this curve can be shown using three parameters: Q_∞, τ_e and α

$$Q(M) = Q_\infty \exp\left[-\left(\frac{\tau_e}{M}\right)^\alpha\right] \tag{4}$$

where Q_∞ is the total specific heat development at M_∞, $Q(M)$ is the specific heat development at maturity M, τ_e is a characteristic time constant and α is a dimensionless curve parameter.

A1.7. Strength development of concrete

The maturity development forms part of the mathematical evaluation of the hardening process; this value is therefore immediately accessible. If requirements are placed on the strength of the concrete at the stripping time, this requirement can be converted into an equivalent maturity requirement based on a measured strength development related to trial castings. Alternatively, the property development of the concrete, including the strength development, may form part of the calculation routines.

The strength properties of the concrete can be documented through casting of cylinders from test mixes. These test cylinders are stored in the mould for 24 h, and are then stored in water at 20°C until testing. At the determination of the strength development, the first tests should be made 8–10 h after casting; the subsequent tests are then to be carried out at logarithmically equidistant intervals (Fig. A.2).

If the strength of the concrete were to be plotted against the maturity, with the maturity plotted logarithmically, the basic form would be an S-shaped curve in Fig. A.2. Analogous to the heat development, this curve can — to a good approximation — be described using the parameters σ_∞, τ_e and α through the expression

$$\sigma(M) \simeq \sigma_\infty \exp\left[-\left(\frac{\tau_e}{M}\right)^\alpha\right] \qquad (5)$$

where σ_∞ is the potential final strength for $M \to \infty$ and $\sigma(M)$ is the strength at maturity M. Expression (5) is purely empirical. Usually the quantities τ_e and α will deviate from the corresponding values for heat development in equation (4).

In the case of concretes with a considerable content of mineral additives with a pozzolanic effect, such as silica fume, a superposed strength increase may sometimes be observed at later ages. This is presumably due to slow, secondary reactions between calcium hydroxide and the added pozzolan. The corresponding feature cannot be seen on the heat development.

A1.8. Criteria for the achievement of resistance against frost

If a hardening concrete freezes before a certain minimum degree of hardening has been achieved, the concrete may be damaged permanently. It is therefore necessary to produce a criterion for when the hardening concrete is frost-resistant. This criterion is expressed in terms of the reaction parameters described in relation to adiabatic heat development.

When it freezes, water will expand by about 9%; consequently, a vital condition for avoiding internal stress within the cement paste during freezing is to have an evenly distributed pore volume corresponding to about 9% of the freezable water.

In a cement paste which is hardening without water access from the surroundings, this pore volume can be formed through the chemical reactions between cement and water, because the products thus formed are of a smaller volume than the reacting cement and water.

It is assumed that only the capillary water is turned into ice during the freezing. The relative volumes of the capillary water V_k and pores V_p during hydration of an originally air-free cement paste can be thus expressed

$$V_k = p - 1{\cdot}4(1-p)R \qquad (6)$$

$$V_p = 0{\cdot}2(1-p)R \qquad (7)$$

where R is the degree of hydration, and p is the porosity of the initial mix (Fig. A.3).

In an originally air-free cement paste which hardens without access to water

1. CASE

CLIENT: ____ CEMENT: ____ JOURNAL NO: ____

BKI-CASE: ____ ADMIXT.: ____ DATED: ____

2. HEAT OF HYDRATION

HEAT CAPACITY: ____ kJ/kg°C

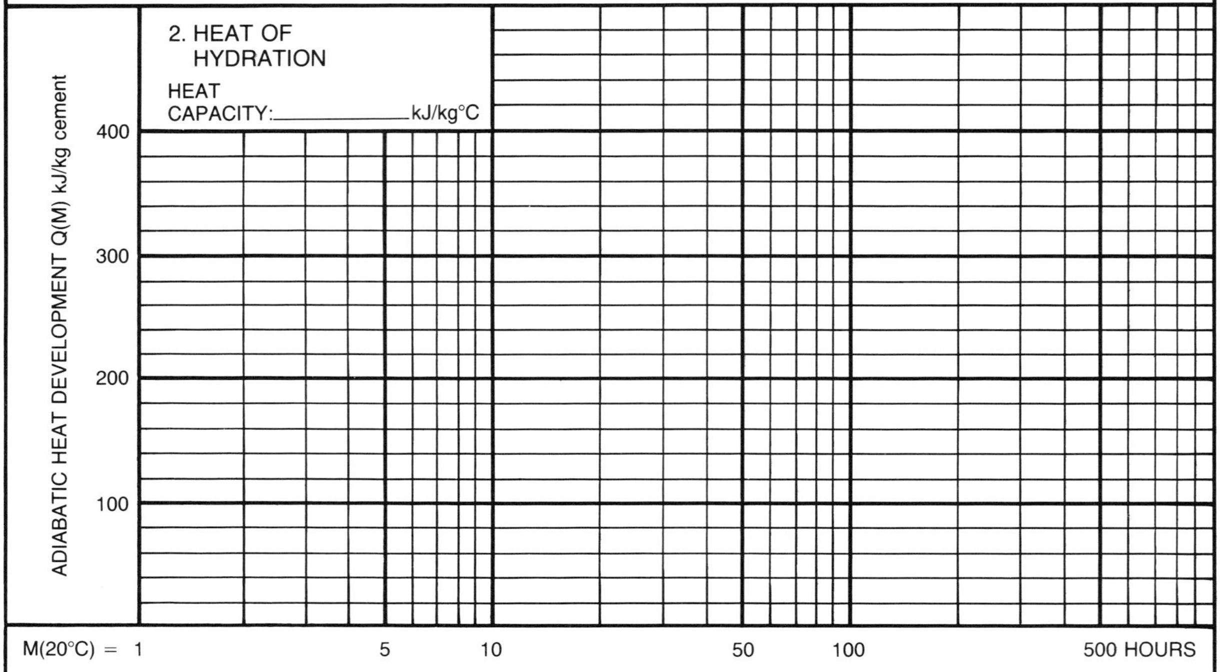

MEASURED: LINEAR: EXPONENTIAL:

3. DESIGN PARAMETERS

LINEAR MODEL: $Q(M) = Q_o \ln (M/\tau_o)$

Q_o = ____ kJ/kg τ_o = ____ h

EXPONENTIAL MODEL: $Q(M) = Q_\infty \exp [-(\tau_e/M)^\alpha]$

Q_∞ = ____ kJ/kg τ_e = ____ h α = ____

4. CONCRETE MIX

	TYPE	DENSITY kg/m^3	WEIGHT kg/m^3	VOLUME m^3	COMMENTS
CEMENT					
WATER					
AIR					
AGGRE-GATE					
ADMIX-TURE					

5. CONCRETE DATA

BULK DENSITY kg/m^3	W/C – RATIO	AIR CONTENT %
MEASURED:	EFFECTIVE:	MEASURED 1:
CALCULATED:	ABSOLUTE:	MEASURED 2:

6.COMMENTS

Date: Init: Revised:

Fig. A.1. Verification sheet for adiabatic calorimetry measurements[36]

1. CASE CLIENT: ________ CEMENT: ________ JOURNAL NO: ________

BKI-CASE: ________ ADMIXT.: ________ DATED: ________

MPa

2. COMPRESSION STRENGTH DEVELOPMENT

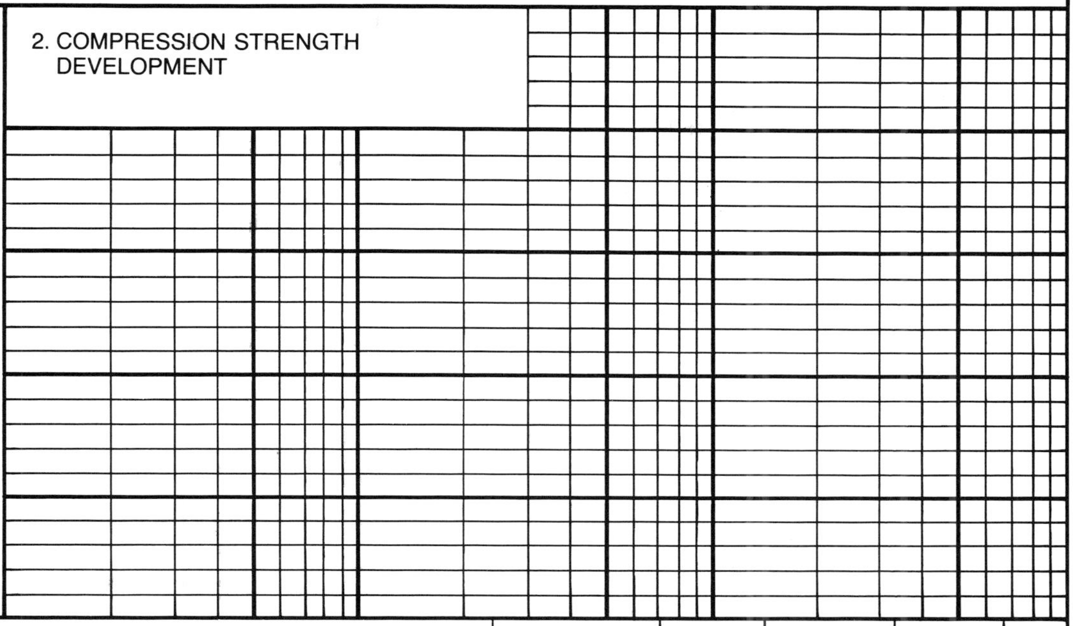

M(20°C) 1 5 10 50 100 500 HOURS

MEASURED: LINEAR: EXPONENTIAL:

3. DESIGN PARAMETERS

LINEAR MODEL: $\sigma(M) = \sigma_o \ln (M/T_o)$

σ_o = ______ MPa T_o = ______ h

EXPONENTIAL MODEL: $\sigma(M) = \sigma_\infty \exp [-(T_e/M)^\alpha]$

σ_∞ = ______ MPa T_e = ______ h α = ______

4. CONCRETE MIX

	TYPE	DENSITY kg/m^3	WEIGHT kg/m^3	VOLUME m^3	COMMENTS
CEMENT					
WATER					
AIR					
AGGRE-GATE					
ADMIX-TURE					

5. CONCRETE DATA

BULK DENSITY kg/m^3	W/C – RATIO	AIR CONTENT %
MEASURED:	EFFECTIVE:	MEASURED 1:
CALCULATED:	ABSOLUTE:	MEASURED 2:

6.COMMENTS

Date: Init: Revised:

Fig. A.2. Verification sheet for strength development measurements[36]

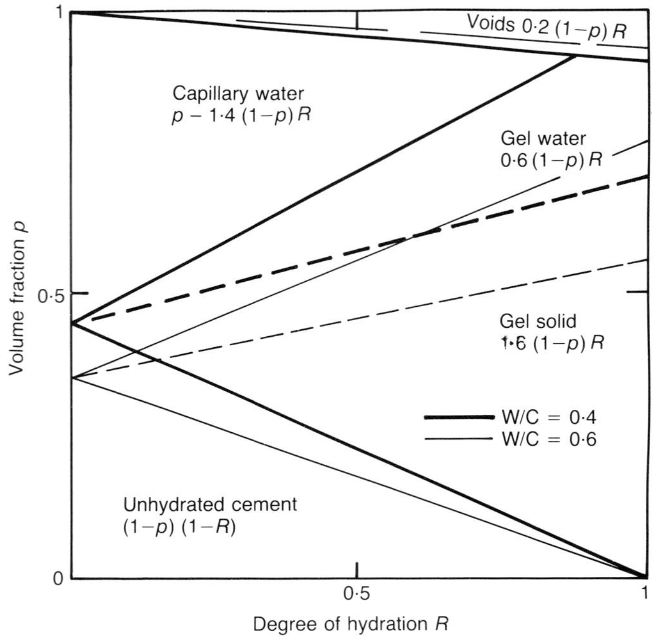

Fig. A.3. Relative volume fractions of the different phases of hardening cement paste (closed system);[38] *(1 − p) = density = 1/(1+3·11(W/C)); p = porosity = 3·11(W/C)/(1+3·11(W/C))*

from the surroundings, the requirement $V_p \geq 0 \cdot 09 V_k$ according to equations (6) and (7) will be met for

$$R \geq 0 \cdot 276 p/(1-p) \tag{8}$$

As the relationship between the p and the W/C ratio for a cement with specific density 3·10 can be given as

$$p/(1-p) = 3 \cdot 10 \text{ (W/C)} \tag{9}$$

the condition (8) can be rewritten as

$$R \geq 0 \cdot 86 \text{ (W/C)} \tag{10}$$

Assuming proportionality between heat development and degree of hardening, the latter can be expressed using the heat development determined by experiment. Thus

$$R \simeq Q(M)/Q_\infty \geq 0 \cdot 86 \text{ (W/C)} \tag{11}$$

If the measured reaction parameters of the adiabatic heat development are inserted, the maturity theoretically necessary for frost resistance during hardening of air-free cement paste without addition of water is found to be

$$M \geq \tau_e/[-\ln(0 \cdot 86 \text{ (W/C)})]^{1/\alpha} \tag{12}$$

In Fig. 10.22 the frost resistance criterion is set against test data from Möller.[39] The broken lines have been calculated for typical measured values of Q_∞, τ_e and α.

A1.9. Criteria for the duration of the curing

One purpose of the curing of concrete is to make sure that the concrete is not exposed to stresses that promote cracks resulting from temperature differences; another is to prevent drying and to make sure that the reaction between cement and water will take place through the whole of the concrete cross-section and provide the hardening intended with the mix proportion of the concrete.

In practice, the curing time will often be a period which the contractor is interested in shortening as much as possible. The following will describe how this curing criterion can be phrased. With a view to planning and control on the site, the time required for curing is expressed in maturity hours (equivalent hardening time at 20°C).

On the assumption of proportionality between the heat development and the degree of reaction, the degree of reaction can be expressed using the parameters resulting from the measured adiabatic heat development

$$R = \frac{Q}{Q_\alpha} = \exp\left[-\left(\frac{\tau_e}{M}\right)^\alpha\right] \tag{13}$$

If this expression is logged

$$M = \tau_e/(-\ln R)^{1/\alpha} \tag{14}$$

The criterion for the curing can then be put into words thus: a suitable part of the theoretically achievable degree of reaction, e.g. no less than 90%, must be achieved, i.e. $R = 0{\cdot}9$.

For W/C ratios of less than 0·45 the theoretical degree of reaction for the closed system will be less than unity, corresponding to the situation where all the capillary water is used in the hydration, resulting in 'self-desiccation'. Low W/C ratios may shorten the duration of the curing. Here the following expression can be set up

$$V_k = p - 1{\cdot}4(1-p)R \tag{15}$$

When $V_k = 0$

$$R = \frac{p}{1{\cdot}4(1-p)} \tag{16}$$

$$\frac{p}{(1-p)} = \frac{\rho_C}{\rho_W}\frac{W}{C} = 3{\cdot}10\,\frac{W}{C} \tag{17}$$

which gives

$$R_{max} = 2{\cdot}21\,\frac{W}{C} \tag{18}$$

$$M_{req} = \frac{\tau_e}{\{-\ln[2{\cdot}21(W/C)]\}^{1/\alpha}} \tag{19}$$

Figure A.4(a) shows the relationship between necessary pre-hardening time for complete hydration and W/C ratio for typical Danish cements. Figure A.4(b) shows the relationship between relative hydration and maturity for the cements, and Fig. A.4(c) shows the relationship between W/C ratio and possible hydration.

For example, suppose a concrete with white Portland cement and W/C = 0·26 is demoulded and exposed to drying after 20 maturity hours. Point A is found for W/C = 0·26 and M = 20 h. Point B is found for white cement and a vertical line through A, and point C is found for W/C = 0·26 and a horizontal line through B.

It can be seen that after 20 h hydration 100% of the theoretically obtainable hydration has taken place. For a W/C ratio of 0·26 this is 58% of complete hydration, i.e. 42% of the cement will never hydrate at this W/C ratio.

A1.10. Drying of fresh concrete

Free capillary water is a vital prerequisite for achieving the desired strength and density during the hardening of the concrete. The evaporation of water from the concrete will take place as from a wet surface until the reaction of the concrete reaches a stage corresponding to 10–20 maturity hours. After this period the water movement within the concrete is guided by diffusion, which is a slow process. It is therefore particularly important to prevent drying during the first 24 h after casting.

The actual quantity of water which may evaporate from a wet concrete surface can be estimated from Figs 10.23 and 10.24.

A1.11. Heat balance in hardening concrete cross-sections

In the centre of thick-walled concrete structures and in highly insulated structures, the heat exchange of the concrete with its surroundings is insignificant compared with the heat generated during the first part of the hardening phase. During this period the temperature increase of the concrete will therefore be approximately proportional to the adiabatic heat development.

In the casting of thin-walled structures, the heat loss to the surroundings will usually be dominant in relation to the hardening heat generated. In order to achieve a reasonable increase in temperature with the consequent rapid strength development, it is necessary to counteract the natural cooling of the concrete in a controlled manner, e.g. by insulating the mould and by introducing curing.

The heat balance to be controlled is sensitive towards changes in the selected level of insulation. This is because

(*a*) if the concrete temperature is lowered, the rate of heat development is reduced
(*b*) if the rate of heat development within the concrete is reduced, the resulting heat balance will often be such that proportionally more heat is given off to the surroundings, so that the change accelerates, and vice versa.

This instability of the heat balance is due to the fact that the heat exchange is proportional to the temperature difference between concrete and air, while the heat development is altered in accordance with the temperature function (equation (1)). If, for example, the concrete temperature is 5°C and the air

Fig. A.4. Maturity, W/C ratio and RH relationships: (a) between maturity and W/C; (b) between maturity and RH; (c) between RH and W/C

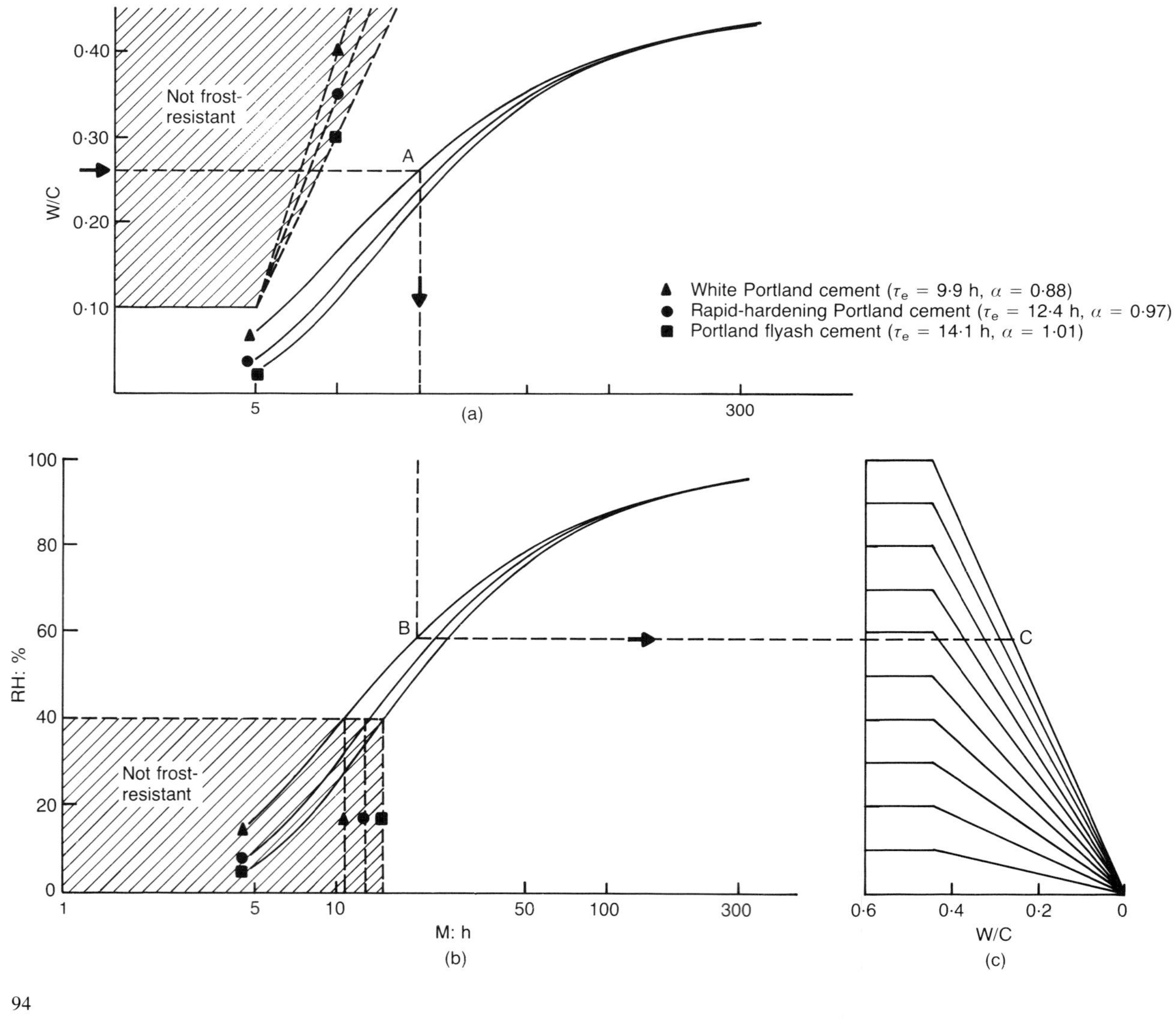

temperature $-10°C$, a change of $1°C$ in the concrete temperature will change the rate of heat development by about 11%, whereas the heat loss will only change by about 6%.

A mathematical calculation of this heat balance problem can in principle be carried out by means of several known numerical methods. The extent of the necessary calculations and — not the least — the possibility of a systematic mapping of the results in a diagram will to a large extent be dependent on the presupposed boundary conditions: the geometry, the adiabatic heat development of the concrete, the starting temperature of the concrete, the air temperature, the wind velocity, and the heat transmission values for mould, cured and free concrete.

A1.12. Coefficient of transmittance

The coefficient of transmittance (COT) is a measure of the degree of insulation applied. In the following, only COTs for convective heat transmission are considered.

The value of the COT is determined by the mould applied, the insulation used, and the convective COT α_k between system and surroundings. The COT α can be determined by calculation

$$\alpha = \left[\frac{1}{\alpha_k} + \left(\frac{e}{\lambda}\right)_{\text{insul}} + \left(\frac{e}{\lambda}\right)_{\text{mould}}\right]^{-1} \text{kJ/m}^2\text{h°C}$$

$$= (m_k + m_{\text{insul}} + m_{\text{mould}})^{-1} \text{ kJ/m}^2\text{h°C} \quad (20)$$

where m_k is the convective coefficient of thermal resistance, λ is thermal conductivity, e is the thickness of the insulation or mould and m is the coefficient of thermal resistance. The convective COT for enforced convection can be approximately calculated as a function of the wind velocity v.

$$\alpha_k \simeq 20 + 14v \text{ kJ/m}^2\text{h°C for } v \leq 5 \text{ m/s} \quad (21)$$

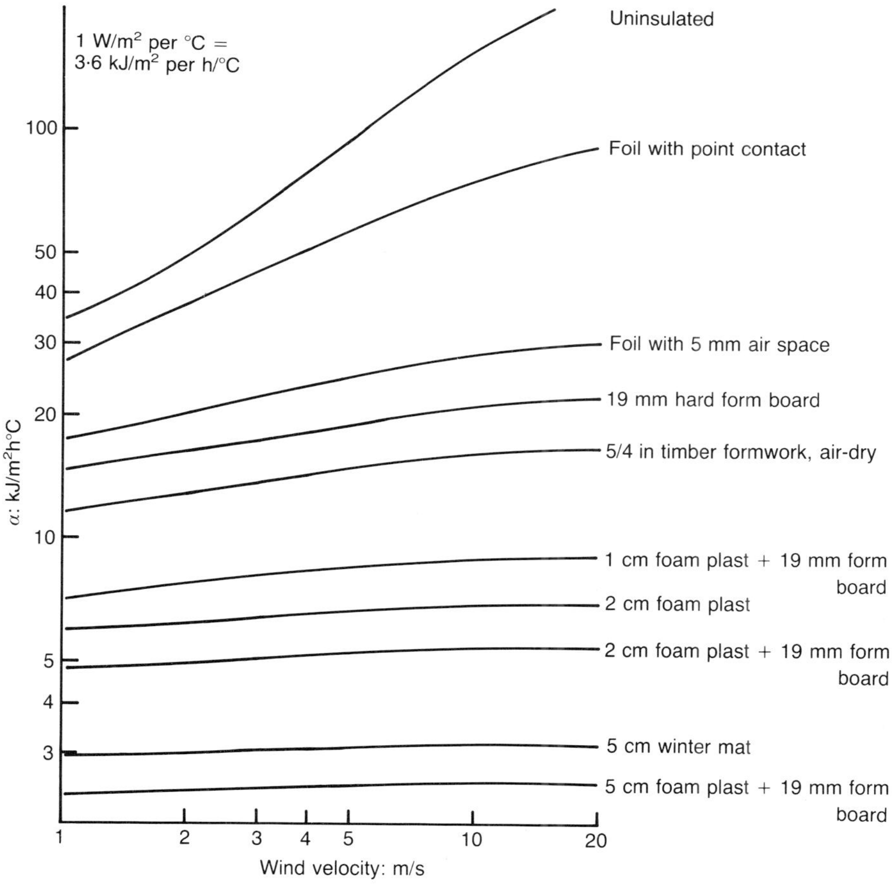

Fig. A.5. Variation of COT with wind velocity for various types of insulation

$$\alpha_k \simeq 25 \cdot 6 v^{0 \cdot 78} \text{ kJ/m}^2\text{h°C for } v > 5 \text{ m/s} \qquad (22)$$

where v is wind velocity at forced heat convection.

Figure A.5 shows calculated values for the COT for a number of commonly used mould and insulation types, plotted against the wind velocity. Both Fig. A.5 and the calculation formulae include only contributions from conduction and convection; contributions from radiation, evaporation, or condensation are not included. Evaporation or condensation of water vapour in connection with heat transmission may have a major effect on the size of the COT. However, in most cases these effects are insignificant for structures in moulds under site conditions.

A1.13. Temperature stress in hardening concrete

With existing knowledge it is impossible to state exact limits to the temperature differences which are acceptable in hardening cross-sections. It is wise to attempt to stay within the following limits for temperature stress

(*a*) for cooling of cross-section after stripping: a maximum difference of 20°C over the cross-section

(*b*) for construction joints and structures with greatly varying cross-section dimensions: a maximum difference of 10–20°C.

Fig. A.6. Example of documentation sheet for asymmetrical cross-section

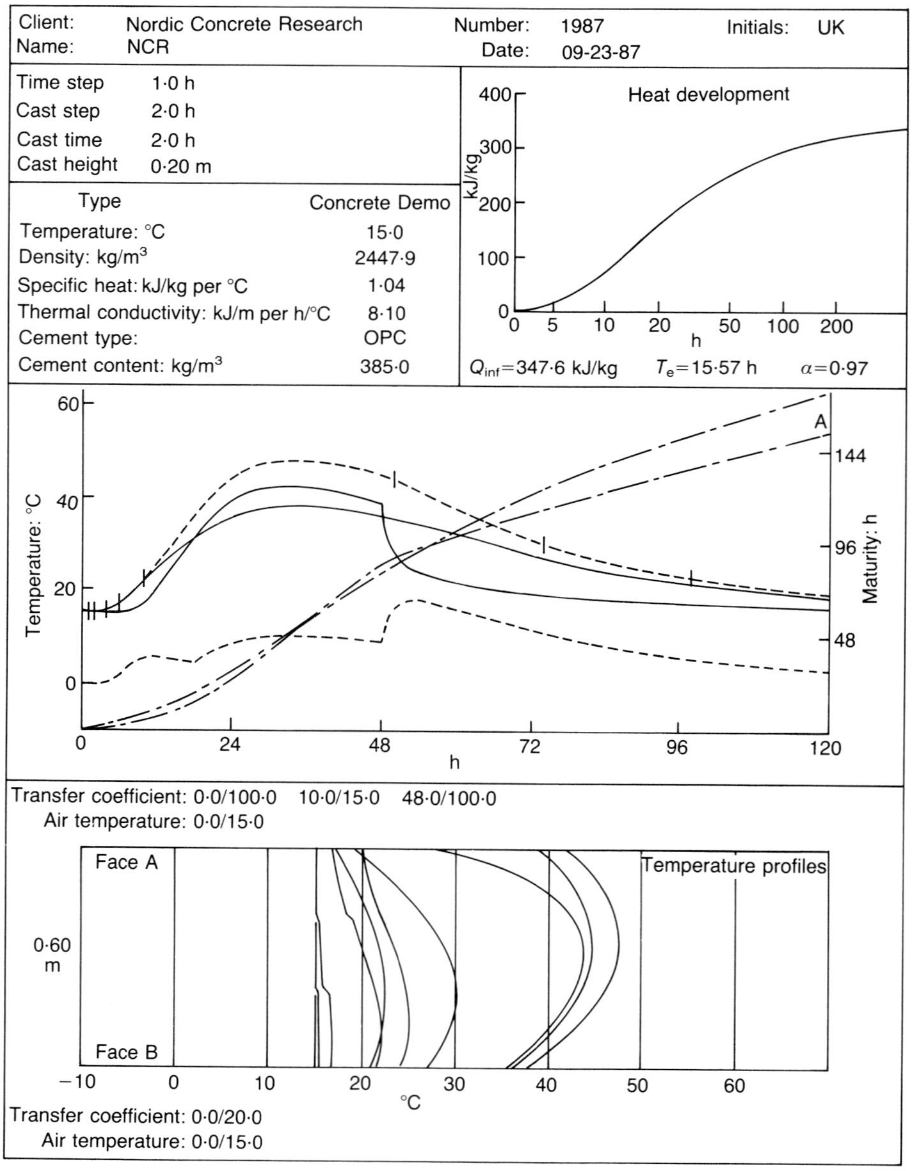

In case (*b*) greater differences in temperature may be acceptable under certain conditions because relaxation effects in the hardening concrete may reduce the stress.

Consequently, in order to avoid thermal cracks it is necessary to be aware of these factors during planning and control of the hardening phase.

A1.14. Planning of construction by computer simulation

A1.14.1. One-dimensional computer simulation

A1.14.1.1. Symmetrical and non-symmetrical cross-sections. Casting of a one-dimensional symmetrical or non-symmetrical plane cross-section (i.e. a wall or a bridge deck) may be planned by computer. The results may be given graphically as shown, for example, in Fig. A.6 as temperature development with time in the centre and at the surface of the cross-section, maturity development with time in the concrete surface and temperature profiles across the section at various times. The non-symmetrical cross-section differs from the symmetrical one only by allowing different boundary conditions on the two concrete surfaces. This results in different maturities at the two surfaces and accordingly both maturities are shown.

A1.14.1.2. Foundations. The casting of fresh concrete against hardened concrete or soil is different from the previous examples in that one concrete surface is cast against a physical structure possessing mass and heat capacitance. Accordingly, a layer of the soil must be included in the calculations.

Often, foundations will be cast in layers with a definite time interval between layers. This possibility must be included in the simulation.

A1.14.2. Two-dimensional and three-dimensional computer simulation
The casting of a concrete structure against an old — hardened — concrete structure (or of, for example, a slab and wall in one pour) can as a start be calculated as separate one-dimensional structures and the temperature development and profiles can be compared, but for an accurate calculation, two-dimensional or three-dimensional computer programs are necessary. These will normally be based on the finite element method, and as such are lengthy and time-consuming. The calculation of temperature profiles is fairly simple and reliable, whereas the calculation of the resulting stresses is more

Fig. A.7 (below left). Example of a wall cast on top of a hardened concrete slab

Fig. A.8 (below right). Example of a wall cast on top of a hardened concrete slab

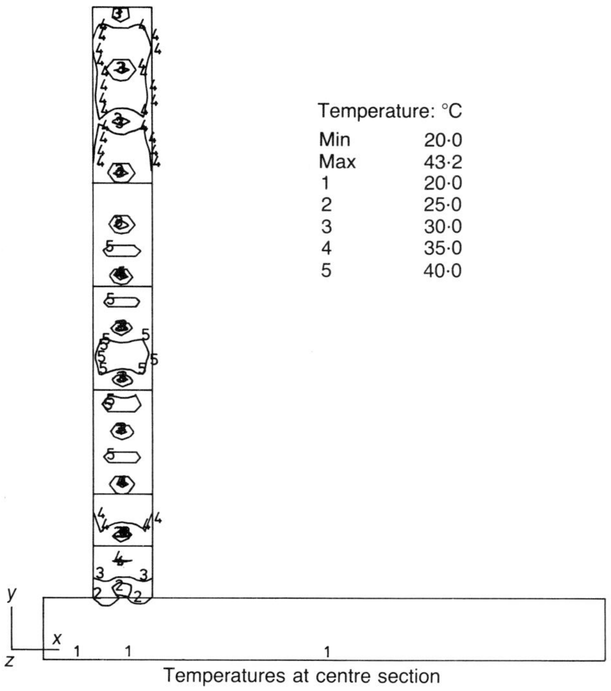

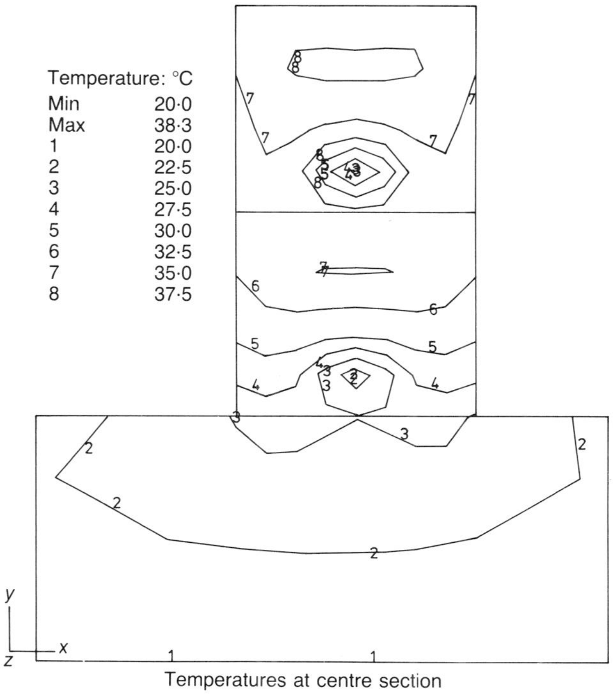

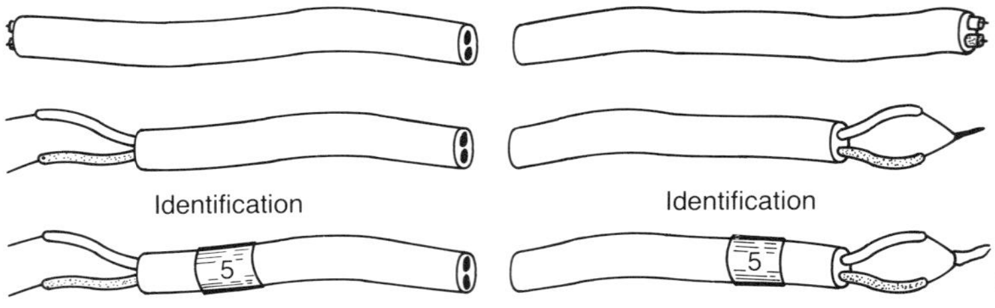

Fig. A.9. Identification, termini and soldering points of thermocouples

uncertain and complicated, because the stiffness and deformability of concrete at an early age change continuously and significantly. Furthermore, the early creep of concrete in a structure may significantly influence the stresses resulting from the temperature differences.

Figures A.7 and A.8 shows examples of the casting of a wall (with cast-in cooling pipes) on top of a hardened concrete slab.

A1.15. Control

Knowledge of the temperature development and temperature distribution in the hardening concrete will give the contractor an opportunity of controlling the hardening phase of the concrete in a safe and appropriate way.

On the basis of temperature measurements it is possible to make the necessary decisions as to stripping, stressing of cables, additional insulation, curing etc. so that prescribed requirements are observed under actual working conditions in an economic way.

In practice it has proved to be appropriate to measure the temperature of the concrete by means of thermocouples of the copper/constantan type, which are placed in the prescribed positions before the casting starts. Figure A.9 shows the fashioning of soldering points at thermocouples. The insulation is removed from the wire ends, and they are twisted and soldered. In principle the measurement can be made with the mix ends twisted but not soldered. However, this involves a risk of failing contact as a result of corrosion of the contact zone at the wire ends. For this reason the wire ends should always

Fig. A.10 (below). Appropriate positioning of measuring points for control of the hardening process of cast concrete structures[40]

Fig. A.11 (right). A number of thermocouples fixed to a bar of isolation material, tied to the reinforcement of a concrete bridge deck[40]

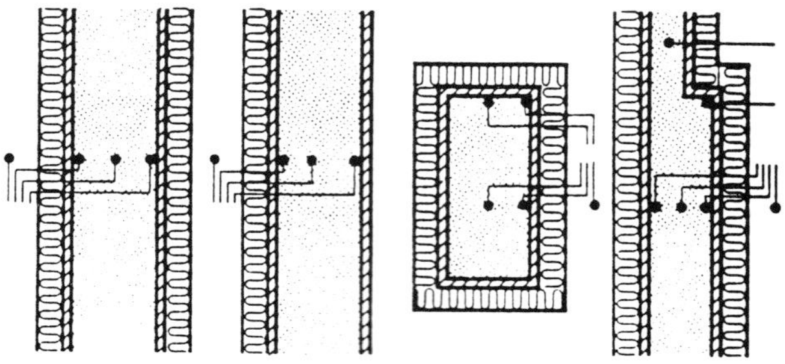

be soldered immediately after de-insulation. Furthermore, a distinct number or colour identification should be used both at the soldering point and at the termini.[40] Figures A.10 and A.11 show examples of thermocouple locations.

The temperature registration may take place either by manual reading of a digital thermometer, in which case the reading is plotted onto a suitable chart, or automatically by means of a data logger from which data can be transferred direct to a computer, where it will be treated mathematically and charted graphically.

It should at all times be possible to follow the temperature and maturity development in a number of selected areas within a concrete structure, so that it is possible to act when the maturity required has been achieved or in order to change the temperature development by altering the insulation capacity of the cover material. The calculation of the degree of maturity is easily done on a chart, using the method described.

The treatment of the measured results will to some extent depend on the purpose of the measuring. Usually, they will aid continuous control of the hardening process, with the aims of

(*a*) prevention against premature freezing
(*b*) observation of stripping time
(*c*) control of maximum temperature
(*d*) control of temperature stress
(*e*) selection of prestressing time for cables
(*f*) selection of additional insulation
(*g*) observation of curing time.

Therefore, it will be appropriate to assess the result after each reading and — if required — to correct the hardening process.

A1.16. The TSSP system for the planning of concreting jobs

The components already discussed — temperature function, adiabatic heat development, strength development, criteria for frost resistance and curing, COT, estimation expressions for temperature differences, response diagrams, and so on — can be applied in an assessment of various phases in the hardening of the concrete. Through computer technology these components can be combined into a coherent tool, as has been done in the Temperature Stress Simulation Program (TSSP).

The TSSP system is based on the systematic application of simulation techniques to hardening control. The system consists of pre-testing of materials, and planning and implementation of casting work together with control of such work. For the user it is a standardized system of printed documentation sheets where the necessary information is printed as a routine. Examples are: documentation sheets with estimated calculation parameters for strength and heat development properties; planning diagrams for casting, comprising typical cross-sections such as walls, columns and stratified castings of foundations on the ground; and two-dimensional analyses of hardening concrete cross-sections. The system also comprises control sheets for use on site.

Figure A.12 shows an example of a planning diagram used to select the final method of casting. It comprises four analyses (apart from a statement of the calculation assumptions): maximum hardening temperature, maximum temperature difference during the hardening, the required mould period to secure a prescribed stripping criterion and an assessment of a possible curing regime. Furthermore, the diagram indicates whether there is a risk of damage to the concrete in the case of early freezing under winter conditions.

The planning diagram may be filled in for related sets of values for casting temperature or for a number of related cross-section dimensions. Therefore, one diagram may contain information from several hundred complete process analyses. On this basis it is possible quickly to form a view of the optimal

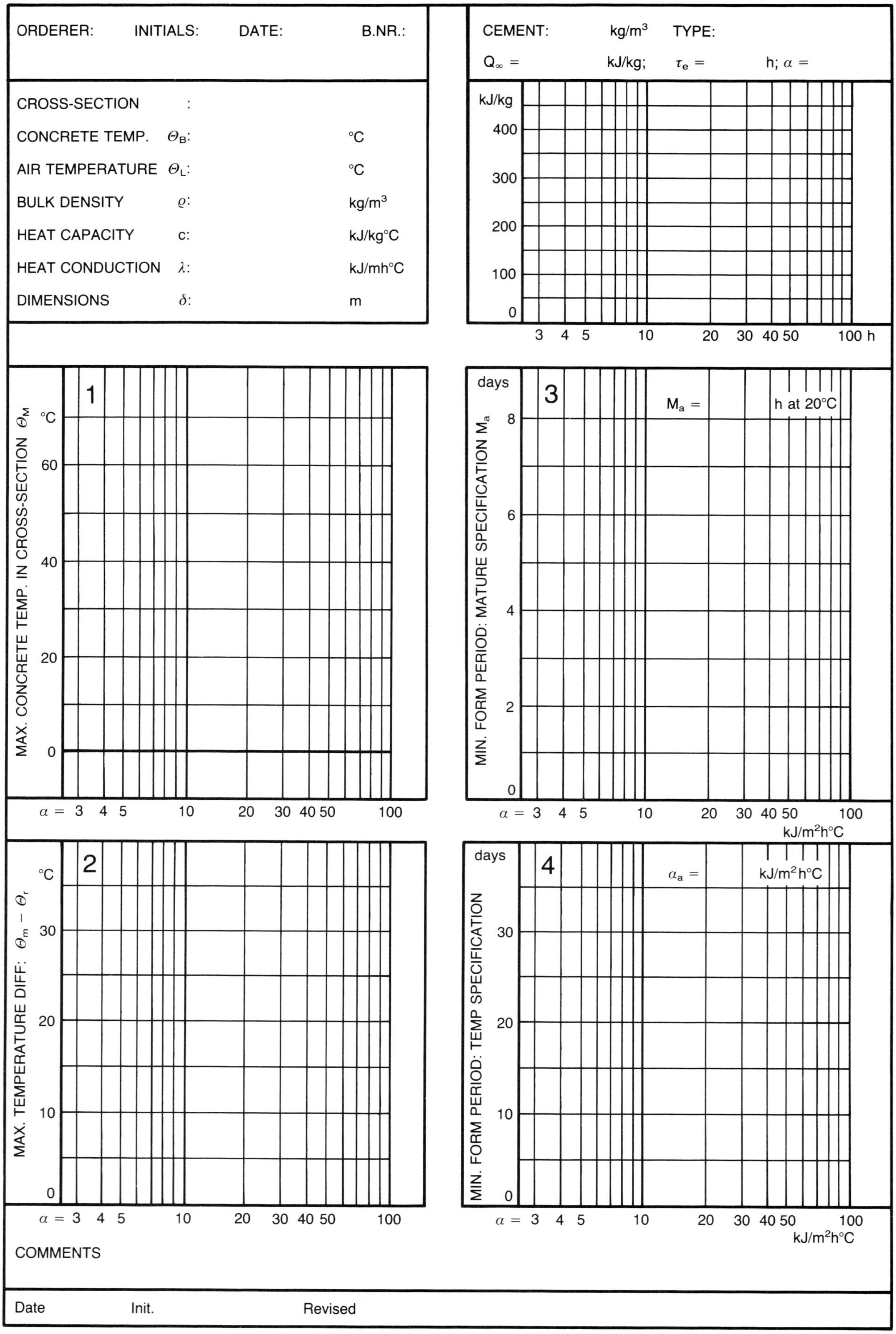

Fig. A.12. Planning diagram form used to select the final method of casting[36]

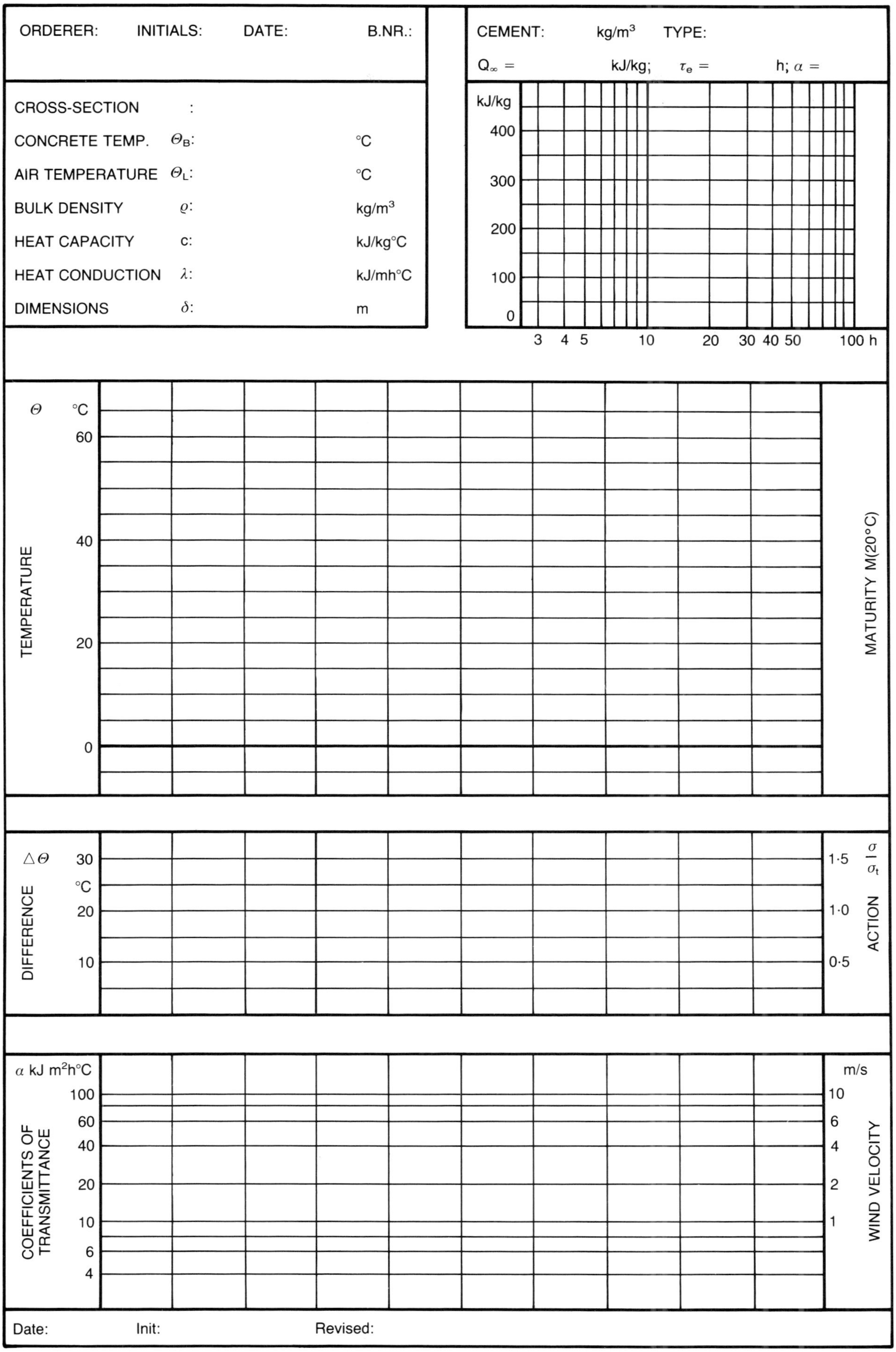

Fig. A.13. Form for plotting details of development of the hardening process[36]

Table A.1. Relative humidity of, for example, curing reinforced concrete pipes

Δp = mmHg	Relative humidity: %		
	Temperature = 10°C	Temperature = 20°C	Temperature = 30°C
2·5	73	86	94
1·0	90	94	97

method in a particular case. The process(es) thus selected may then be plotted separately for closer evaluation.

Figure A.13 shows a similar form for plotting details of the hardening process.

A1.17. Practical examples

The following examples are reproduced from ref. 36.

A1.17.1. Rate of evaporation

A newly-cast concrete pipe is stored in a production plant with an air temperature of 25°C and a humidity of 70%. The temperature of the concrete in the pipe wall is 27°C. The rate of evaporation is sought for a wind velocity of 2 m/s.

The water film on the concrete surface has a temperature of 27°C. In the boundary layer the relative humidity, RH = 100%. The point A (27°C, 100%) on Fig. 10.23 indicates that the vapour pressure p_1 = 27 mmHg on the surface. The ambient air temperature = 25°C and RH = 70%. The reading for point B gives p_2 = 16·5 mmHg. The vapour pressure difference Δp = 27·0 − 16·5 = 10·5 mmHg. Putting this into Fig. 10.24 for v = 2 m/s (point C), an evaporation loss $\dot{W}$ ~ 0·4 kg/m²h is obtained.

A1.17.2. Curing conditions for reinforced concrete pipes

A production of reinforced concrete pipes is to be cured for 24 h in a factory. The curing conditions have been specified as follows. The drying shrinkage in the setting period 0–8 h after concreting is to be reduced by appropriate measures. During this period the potential evaporation (i.e. the evaporation from the wet surface at ambient temperature) should constitute less than 5% of the water content of the concrete. What curing conditions should be ensured in the factory to satisfy the specifications given for reinforced pipes with a wall thickness of 60 mm?

If the evaporation loss from the interior surface of the pipes is assumed to be 50% of the loss from the outside, the permissible evaporation loss $\dot{W}$ for concrete with a water content of about 135 l/m³ is found from

$$\dot{W} \approx (0 \cdot 05 \times 135 \times 0 \cdot 06/1 \cdot 5)/8 \approx 0 \cdot 034 \text{ kg/m}^2\text{h}$$

From readings for v = 0 m/s and 2 m/s the permissible vapour pressure differences are found to be ~2·5 and 1 mmHg, respectively.

From the vapour pressure diagram the corresponding relative humidity RH is found for the relevant temperature (Table A.1). To satisfy the specifications, the relative humidity in the factory should be kept within the limits shown in Fig. A.14. In order to limit the evaporation loss it is thus important to protect the concrete carefully against draught, for example from open gates or doors in the factory. If the curing temperature is increased, stricter control is also necessary for keeping the relative humidity in the factory at a high level — which is known from experience to cause a lot of problems.

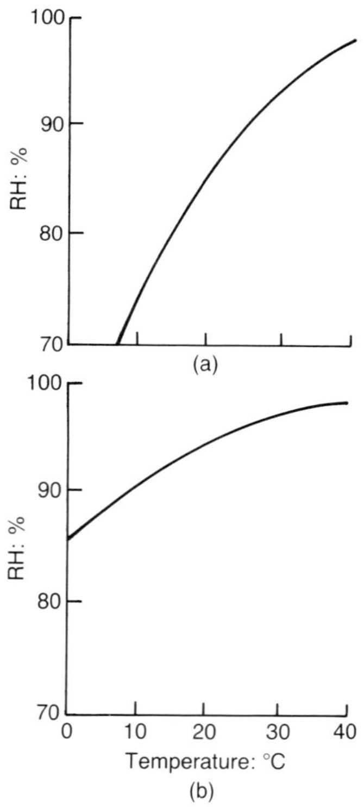

Fig. A.14. Relative humidity requirements for example of reinforced concrete pipes: (a) v = 0 m/s; (b) v = 2 m/s

A1.17.3. Comparison of climatic conditions

A rough comparison is desired between the potential evaporation during the day in Copenhagen and Kuwait, using the following climatic data for summer conditions: Copenhagen, temperature approximately 21 °C, RH about 70%; Kuwait City, temperature approximately 46 °C, RH about 25%. Wind velocity is assumed to be 5 m/s in both cases.

The wanted vapour pressure differences are obtained as follows. For Copenhagen: $\Delta p = p(21°C, 100\%) - p(21°C, 70\%) \sim 19 - 13 = 6$ mmHg. For Kuwait City: $\Delta p = p(46°C, 100\%) - p(46°C, 25\%) \sim 76 - 18 = 58$ mmHg. Using these figures, one finds from Fig. 10.24 for Copenhagen that $\dot{W} \sim 0{\cdot}42$ kg/m^{2}h. For Kuwait City the entry values lie beyond the diagram axes. Using the fact that $\dot{W}$ is proportional to p it is found that $\dot{W} \sim 10\dot{W}(58/10$ mmHg, 5 m/s$) \sim 10 \times 0{\cdot}41 = 4{\cdot}1$ kg/m^{2}h.

In Copenhagen the potential evaporation during the day will thus be $\sim 0{\cdot}4$ kg/m^{2}h in July. In Kuwait City a rate of evaporation approximately ten times higher must be assumed.

A1.17.4. Estimation of curing chamber conditions

Due to problems with drying-out of concrete during floor curing of pavement blocks, a factory has decided to establish a curing chamber for heat curing of the concrete. Comparison between existing and expected conditions of manufacture has given

(*a*) floor curing, summer: $\theta \sim 20$°C, RH $\sim 50\%$, $v \sim 0{\cdot}5$ m/s
(*b*) floor curing, winter: $\theta \sim 15$°C, RH $\sim 80\%$, $v \sim 0{\cdot}5$ m/s
(*c*) chamber curing: $\theta \sim 50$°C, RH $\sim 95\%$, $v \sim 4{\cdot}0$ m/s.

Will the change to chamber curing reduce the drying-out problem? Condensation, heat of evaporation and heat of hydration are disregarded.

Comparison between potential evaporation under the existing and the expected conditions of manufacture gives

(*a*) floor curing, summer: $\dot{W}(20°C, 50\%, 0{\cdot}5$ m/s$) \sim 0{\cdot}2$ kg/m^{2}h
(*b*) floor curing, winter: $\dot{W}(15°C, 80\%, 0{\cdot}5$ m/s$) \sim 0{\cdot}1$ kg/m^{2}h
(*c*) chamber curing: $\dot{W}(50°C, 95\%, 4{\cdot}0$ m/s$) \sim 0{\cdot}4$ kg/m^{2}h.

It can then be expected that the maximum drying-out rate will occur during chamber curing.

A1.17.5. Plastic foil covering of slabs

In the efforts to reduce evaporation from newly cast slabs, tests have been made with plastic covering. During the test air velocity, humidity, and temperature were measured as shown in Fig. A.15.

These measurements give

(*a*) stack A: $\Delta p(20{\cdot}5°C, 100\%, 20°C, 78\%) = 4{\cdot}4$ mmHg;
$\dot{W}(v = 0{\cdot}5$ m/s$) = 0{\cdot}9$ kg/m^{2}h
(*b*) stack B: $\Delta p(26{\cdot}0°C, 100\%, 22°C, 100\%) = 5{\cdot}5$ mmHg;
$\dot{W}(v = 0$ m/s$) = 0{\cdot}8$ kg/m^{2}h.

It can be seen that the rate of evaporation has not been reduced significantly by the plastic covering. The water evaporates from the surface of the fresh slabs and is condensed on the plastic foil and drained away. To prevent this loss of water the foil should be in contact with the concrete surfaces.

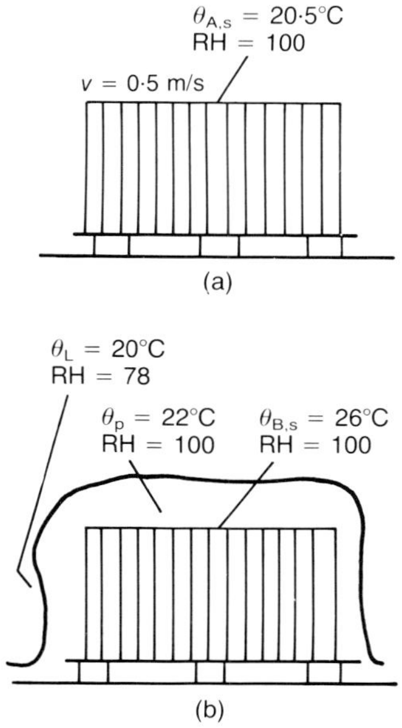

Fig. A.15. Air velocity, temperature and humidity for newly-cast slabs: (a) open; (b) loosely covered with plastic foil

A1.17.6. Estimation of temperature differences during hardening

A solid 80 cm thick ribbed beam reaches a maximum temperature of 74 °C. During hardening the temperature of the ambient air is 8 °C. The concrete is cast in an insulated mould with coefficient of transmission $\alpha = 2{\cdot}5$ kJ/m^{2}h°C and a wind velocity of 5–6 m/s. Ordinary cement is used. An estimate of the maximum temperature difference between the middle of the cross-section and the boundary is desired.

Assuming that cooling is symmetrical, we obtain $\alpha = 2{\cdot}5$ kJ/m^{2}h°C, $\delta = 0{\cdot}4$ m and $\theta_c - \theta_a = 66$°C. Using these values in Fig. 10.25, a maximum temperature difference over the cross-section during hardening of about 4°C is predicted.

A1.17.7. Choice of supplementary insulation

The ribbed beam dealt with in the example above is stripped when the maximum temperature θ_c = 74°C. In this connection supplementary insulation must be chosen to ensure that the specification of maximum temperature difference of 20°C over the cross-section is complied with.

Assuming symmetrical cooling: $\theta_c - \theta_s$ = 20°C, $\theta_c - \theta_a$ = 66°C and δ = 0·4 m.

Using these values in Fig. 10.25 starting at the point $(\theta_c - \theta_s)$, $(\theta_c - \theta_a)$ = (20,66) results in the choice of supplementary insulation with $\alpha \simeq$ 17 kJ/m^{2}h°C. This ensures the fastest cooling of the cross-section.

A1.17.8. Termination of supplementary insulation

For the cross-section dealt with above, the supplementary insulation used may be removed when $(\theta_c - \theta_a)$ has decreased to a value that ensures that specifications for a maximum temperature difference of 20°C is complied with for a free, unprotected concrete surface.

At a wind velocity v of 5–6 m/s the convective coefficient of thermal transmittance α will be about 100 kJ/m^{2}h°C. Using the dimension dealt with above, δ = 0·4 m, in the diagram starting at $(\theta_c - \theta_s)$ = 20°C it is found that cooling with the chosen insulation must be continued until $\theta_c - \theta_a \leq$ 27°C. At θ_a = 8°C as assumed this means that the insulation should not be removed until $\theta_c \leq$ 35°C.

A1.17.9. Changes in supplementary insulation procedures

A column of characteristic dimension R = 0·8 m is cast. The casting is planned to occur at a time when the typical value for air temperature θ_a = 10°C and wind velocity v = 2 m/s. Concrete temperature is estimated to rise 40°C compared with the temperature at casting, which is 15°C. The form system may be stripped when concrete temperature has reached θ_c = 55°C. Based on these assumptions, supplementary insulation has been bought: tarpaulin placed properly on wooden beams may result in a coefficient of thermal transmittance $\alpha \sim$ 16 kJ/m^{2}h°C.

However, the casting is postponed and immediately before it starts the weather changes. It becomes windy, with wind velocity 5 m/s and the air temperature is lowered to θ_a = −10°C. Using these altered assumptions, the coefficient of thermal transmittance should be less than about 9 kJ/m^{2}h°C. The corresponding level of insulation can be attained by using insulation at intervals, as shown in Fig. 10.25.

References

1. Comité Euro-International du Béton. *Durability of concrete structures*. CEB, Paris, 1982, State of the art report, Bulletin d'Information 148.
2. Comité Euro-International du Béton and RILEM. *Durability of concrete structures*. CEB, Lausanne, 1984, International workshop 1983 report, Bulletin d'Information 152.
3. Comité Euro-International du Béton. *Draft CEB guide to durable concrete structures*. CEB, Lausanne, 1985, Bulletin d'Information 166.
4. Comité Euro-International du Béton. Second international workshop with RILEM, Bologna, 1986. Unpublished.
5. Concrete Society. *Non-structural cracks in concrete*. Concrete Society, 1982, Technical report 22.
6. Comité Euro-International du Béton. *CEB-FIP model code for concrete structures*. CEB, Paris, 1978, Bulletin d'Information 124/125.
7. Comité Euro-International du Béton. *Design manual on cracking and deformations*. CEB, Lausanne, 1985, Bulletin d'Information 158.
8. American Society for Testing and Materials. *Limits on aluminates in concrete for sulphate resistance*. ASTM, Philadelphia.
9. Thistlethwayte D.K.B. *The control of sulphides in sewerage systems*. Butterworths, 1972. Translated as *Sulphide in Abwasseranlagen, Ursachen, Auswirkungen, Gegenmassnahmen*. Beton-Verlag, 1979.
10. Powers T.C. *Properties of fresh concrete*. Wiley, New York, 1968.
11. Comité Européen de Normalisation. *Concrete — performance, production, placing and compliance criteria*. CEN, 1984, Draft document prEN 206.
12. British Standards Institution. *Code of practice for the design of steel bridges*. BSI, London, 1982, BS 5400.
13. British Standards Institution. *Code for agricultural building*. BSI, London, BS 5502.
14. De Sitter. Costs for service life optimization. In *Durability of concrete structures*. CEB, Lausanne, 1984, International workshop 1983 report, Bulletin d'Information 152.
15. Cembureau. *Use of concrete in aggressive environments*. Cembureau, 1978, Recommendations.
16. Regourd M. Durability. Physico-chemical and biological processes related to concrete. In *Durability of concrete structures*. CEB, Lausanne, 1984, International workshop 1983 report, Bulletin d'Information 152.
17. Jessen J.J. Construction and maintenance for improved durability. In *Durability of concrete structures*. CEB, Lausanne, 1984, International workshop 1983 report, Bulletin d'Information 152.
18. Comité Euro-International du Béton. *Detailing of concrete structures*. CEB, Paris, 1982, Bulletin d'Information 150.
19. Comité Euro-International du Béton. *Industrialization of reinforcement in reinforced concrete structures*. CEB, Lausanne, 1985, Bulletin d'Information 164.
20. Huberty J.M. *Durabilité d'aspect des bétons apparants — le vieillissement des facades*. Centre Scientifique et Technique de la Construction and Federations of the Cement and Concrete Industries, Brussels, 1980.
21. Hawes F. *The weathering of concrete buildings*. Cembureau.
22. International Organization for Standardization. *Concrete — determination of compressive strength of test specimens*. ISO, 1978, ISO 4012.
23. International Organization for Standardization. *Concrete — determination of air content of freshly mixed concrete — pressure method*. ISO, 1980, ISO 4848.
24. International Organization for Standardization. *Concrete hardened — determination of the depth of penetration of water under pressure*. ISO, 1983, Draft standard 7031.
25. International Organization for Standardization. *Concrete — determination of scaling resistance of surfaces exposed to de-icing chemicals*. ISO, 1984, Draft standard 4846.
26. Biczok I. *Concrete corrosion — concrete protection*, 8th edn. Akademiai Kiado, Budapest, 1972.
27. American Society for Testing and Materials. *Specification for concrete aggregates*. ASTM, Philadelphia, 1990, C33.

28. American Society for Testing and Materials. *Practice for petrographic examination of aggregates for concrete*. ASTM, Philadelphia, 1985, C295.
29. American Society for Testing and Materials. *Test method for potential reactivity of aggregates (chemical method)*. ASTM, Philadelphia, 1987, C289.
30. American Society for Testing and Materials. *Test method for potential alkali reactivity of cement-aggregate combinations (mortar-bar method)*. ASTM, Philadelphia, 1987, C227.
31. American Society for Testing and Materials. *Test method for potential alkali reactivity of carbonate rocks for concrete aggregates (rock cylinder method)*. ASTM, Philadelphia, 1969, C586.
32. Comité Euro-International du Béton. *CEB-FIP model code 1990 — first predraft 1988*. CEB, Lausanne, 1988, Bulletin d'Information 190a/190b.
33. Comité Euro-International du Béton. *Diagnosis and assessment of concrete structures*. CEB, Lausanne, 1989, State of the art report, Bulletin d'Information 192.
34. Comité Euro-International du Béton. Redesign of concrete structures. Unpublished draft report.
35. American Society of Civil Engineers. *Guide to investigation of structural failures*. ASCE, New York, 1979.
36. Beton- og Konstruktionsinstituttet. *Dokumentationsblade*. 1978–84.
37. Cementfabrikkernes tekniske Oplysningskontor CtO. *Beton-Teknik*, 1981, Oct. Betons Hærdevarme.
38. Freisleben-Hansen P. *Hærdeteknologi. 1: Portlandcement*. Aalborg Portland og BFK-centralen, 1978.
39. Möller G. *Materialproblem vid vinterbetongarbetan, tidig frysning av betong*. Svenska Forskningsinstituttet för cement och beton, Stockholm, 1962, Utredninger 5.
40. Statens Byggeforskningsinstitut. *Vinterstøbning af beton m. tillæg*. Arbejdsblok med beregnings- og kontrolskema til brug ved winterstøbning af beton, Copenhagen, 1982.

Bibliography

The references listed in this section have been selected from the enormous amount of available literature treating the different aspects of the durability problem. The list does not pretend to be complete but covers the major subjects relevant to a comprehensive survey on contemporary research and investigations into the durability of concrete structures.

1945 Pourbaix M. Thesis, Delft.

1946 Jackson F. H. The durability of concrete in service. *J. Am. Concr. Inst.*, **18**, No. 2, Oct.

1951 Holmberg Å. Two highway bridges with high-grade steel reinforcement. *Publs Int. Ass. Bridge Struct. Engng*, No. 11.

1956 Klas H. and Steinrath H. *Die Korrosion des Eisens und ihre Verhütung*. Verlag Stahleisen, Dusseldorf.

1957 Holmberg Å. Investigations on cracked concrete structures. *Symposium on bond and crack formation in reinforced concrete*. RILEM.

1958 Legget R. F. and Hutcheon N. B. The durability of buildings. *Symposium on some approaches to durability in structures, Boston*. American Society for Testing and Materials, Philadelphia, SP 236.

1960 Evans U. R. *The corrosion and oxidation of metals*. Arnold, London.

1964 Moll H. L. Über die Korrosion von Stahl in Beton. *Dt. Aussch. Stahlbet.*, No. 169.
Rehm G. and Moll H. L. Versuche zum Studium des Einflusses der Rissbreite auf

die Rostbildung an der Bewehrung von Stahlbetonbauteilen. *Dt. Aussch. Stahlbet.*, No. 169.

1965 Kleinschmidt H. J. Untersuchung über das Fortschreiten der Carbonatisierung an Betonbauwerken Tiefe der carbonatisierten Schicht alter Betonbauten — Untersuchung an Betonproben. *SchrReihe Dt. Aussch. Stahlbet.*, No. 170.

Kungl. Väg- och Vattenbyggnadsstyrelsen (Board of Roads and Waterways). *Undersökninger av broar med avseende på sprickbildning* (A study on cracks in bridges). KVV, Stockholm, K568:2.

Rehm G. and Moll H. L. Beobachtungen an alten Stahlbetonbauteilen hinsichtlich Carbonatisierung des Betons und Rostbildung an der Bewerhung. *Schr Reihe Dt. Aussch. Stahlbet.*, No. 170.

1966 Kaesche H. *Die Korrosion der Metalle*. Springer-Verlag, Berlin, Heidelberg, New York, 1966.

Plum N. M. *et al.* A new approach to testing of building materials. Fragmentary contributions to the philosophy of testing. *R.I.L.E.M. Bull.*, No. 30, Mar.

Powers T. C. Comments on A new approach to testing of building materials, by N.M. Plum *et al.* (*R.I.L.E.M. Bull.*, No. 30), *R.I.L.E.M. Bull.*, No. 33, Dec.

1967 Mather B. *Concrete deterioration*. U.S. Army Engineer Waterways Experiment Station, Corps of Engineers, Miscellaneous Paper 6-927.

Meyer A. *et al.* Karbonatisierung von Schwerbeton. Einfluss von Luftkohlensaure und Feuchtigkeit auf die Beschaffenheit des Betons als Korrosionsschutz für Stahleinlagen. *SchrReihe Dt. Aussch. Stahlbet.*, No. 182.

1968 Biczok I. *Betonkorrosion — Betonschutz*. Bauverlag, Wiesbaden, Berlin.

Lauer K. R. *et al. A durability study of concrete using Monte Carlo simulation*. University of Notre Dame, Indiana, Supplementary paper iii-39.

Liesche H. and Paschke K. *Beton in agressiven Wassern*. Berlin.

Locher F. W. Untersuchung des Betons von Üferschutzbauten auf Helgoland. *Beton* **18**, No. 2, 47–50; No. 3, 82–84.

Rehm G. and Rauen A. Korrosion von Stahl in Beton. *Betonwerk Fertigteil-Technik*, No. 5, 258–264.

Woods H. *Durability of concrete construction*. American Concrete Institute, Detroit, Monograph 4.

Zimbelmann R. Über das Verhalten von zementstein und Beton bei niederen Temperaturen. Teil I und II. Otto-Graf-Institut, *SchrReihe*, Nos 67–68.

1969 Nawy E. G. and Orenstein G. S. Crack width control in two-way concrete slabs reinforced with welded wire fabric. *Engng Res. Bull.*, No. 47. Rutgers University, New Jersey.

Richartz W. Die Bindung von Clorid bei der Zementerhartung. *Zem.-Kalk-Gips*, No. 10, 447–456.

Walz K. and Wischers G. Über den Widerstand von Beton gegen die mechanische Einwirkung von Wasser höher Geschwindigkeit. *Betontech. Ber.*, No. 9, 403–405; No. 10, 457–460.

1970 Kordina K. and Waubke N. V. Über den Erhaltungszustand 20 Jahre alter Spannbetontrager. *SchrReihe Dt. Aussch. Stahlbet.*, No. 212.

Locher W. and Sprung S. Einwirkung von salzsaurehaltigen PVC — Brandgasen auf Beton. *Beton*, No. 2, 63–65; No. 3, 99–104.

1971 Beeby A. W. *An investigation of cracking on the side faces of beams*. Cement and Concrete Association, Wexham Springs, Technical Report 42.466.

Nawy E. G. and Blair K. *Further studies on flexural crack control in structural slab system cracking, deflection and ultimate load of concrete slab systems*. American Concrete Institute, Detroit, SP30-1.

Nawy E. G. and Ornstein G. S. Crack width control in reinforced concrete two-way slabs. *J. Struct. Div. Am. Soc. Civ. Engrs*, 701–721; 1804.

1972 Biczok I. *Concrete corrosion — concrete protection*, 8th edn. Akademiai Kiado, Budapest.

Houston J. T. *et al. Corrosion of reinforcing steel embedded in structural concrete*. University of Texas, Research Report 112-1-F.

Kommission Carbonatisierung. Carbonatisierung des Betons — einflusse und auswirkungen auf den Korrosionsschutz der Bewehrung. *Beton* 22, No. 7, 296–299.

1973 Atimtay E. and Ferguson P. M. Early chloride corrosion of reinforced concrete — a test report. *J. Am. Concr. Inst.*, No. 9, 606–611.

Beeby A. W. *Suggested modifications to the crack prediction formula in the 1970 C.E.B. Recommendations*. CEB, Paris, Bulletin 89.

Weigler H. and Segmuller E. Einwirkung von Chloriden auf Beton. *Betonwerk Fertigteil-Technik*, No. 8, 577–584.

1974 Blevot J. Pathologie des constructions en beton armé. *Annls Inst. Tech. Bâtim.*, No. 320, Sept.; Series: Gros Oeuvre, No. 23.

Klas H. and Steinrath H. *Die Korrosion des Eisens und ihre Verhutung*. Verlag Stahleisen, Dusseldorf.

Martin H. Zeitlicher Verlauf der Chloridwanderung im Beton, der einem PVC — Brand ausgesetzt war. *Betonwerk Fertigteil-Technik*, No. 1, 19–24; No. 2, 89–95.

1975 American Concrete Institute. *Durability of concrete*. ACI, Detroit, SP 47.

Behaviour in service of concrete structures. Inter-Association Colloquium, Liège, Preliminary reports, vols I–III, Final report.

CUR. *Carbonatatie Lichtbeton. Literaturstudie*. CUR.

Les constructions en béton armé. *Pathologie de la construction*, I–III. Batir nos 33–39.

Martin H. *et al.* Carbonation of concrete made with different types of cement. *Behaviour in service of concrete structures*. Inter-Association Colloquium, Liège, Preliminary report, vol. III.

Martin H. *et al.* Karbonatisierung von Beton aus verschiedenen Zementen. *Betonwerk Fertigteil-Technik*, No. 12, 588–590.

Neville A. M. *Properties of concrete*. Pitman, London.

Schiessl P. Admissible crack width in reinforced concrete structures. *Behaviour in service of concrete structures*. Inter-Association Colloquium, Liège, Preliminary report, vol. III.

1976 American National Standards Institute and American Society for Testing and Materials. *Standard definitions of terms relating to hydraulic cement*. ANSI/ASTM, C219-76a.

Buhrer R. *et al.* Untersuchungen an 20 Jahre alten Spannbetontragern. *SchrReihe Dt. Aussch. Stahlbet.*, No. 271.

FIP. *Report on prestressing steel. 1. Types and properties*. FIP, 5/3.

Heitz E. and Schwenk W. Theoretische Grundlagen der Ermittlung von Korrosionsstromdichten aus Polarisationswiderstanden. *Werkstoffe Korros.*, **27**, 241–245.

Kalousek G. L. *et al.* Past, present and potential developments of sulfate-resisting concretes. *J. Test. Evaluation*, **4**, No. 5, Sept.

Manning D. G. and Ryell J. *Durable bridge decks*. Ministry of Transportation and Communications, Downsview, Ont., RR 203.

Organisation for Economic Co-operation and Development Road Research Group. *Bridge inspection*. OECD, Paris.

Schiessl P. Zur Frage der zulassigen Rissbreite und der erforderlichen Betondeckung im Stahlbetonbau unter besonderer Berücksichtigung der Karbonatisierung des betons. *Dt. Aussch. Stahlbet.*, No. 255.

Springenschmid R. Über die Dauerhaftigkeit von Bauwerken aus Beton und Stahlbeton. *Zem. Bet.*, No. 5, 224–229.

1977 American Concrete Institute Committee 201. Guide to durable concrete. *J. Am. Concr. Inst.*, Dec.

American National Standards Institute and American Society for Testing and Materials. *Standard terminology relating to erosion and wear*. ANSI/ASTM, G40-77.

American Society for Testing and Materials. *Chloride corrosion of steel in concrete*. ASTM, Philadelphia, STP 629.

Arup H. Galvanic action of steel in concrete. *Korrosionscentralen*, Aug.

Bonzel J. and Siebel E. Neuere Untersuchungen über den Frost-Tausalz-Widerstand von Beton. *Beton*, 4–6.

Kern E. Korrosionsschutz von Stahl im Beton. *VDI-Ber.*, No. 285, 83–88.
Ludwig U. and Darr G. M. *Über die Sulfatbestandigkeit von Zementmortel.* Westdeutscher Verlag, Opladen.

1978 American Concrete Institute. *Corrosion of metals in concrete.* ACI, Detroit, SP 49.
American Society for Testing and Materials. Durability of building materials and components. *Proc. 1st Int. Conf. Bldg Mater. Ottawa.* ASTM, Philadelphia, STP 691.
Andrade C. and Gonzales J. A. Quantitative measurements of corrosion rate of reinforcing steels embedded in concrete using polarization resistance measurements. *Werkstoffe Korros.*, **29**, 515–519.
Bazant Z. P. *Physical model for corrosion of steel in concrete.* CBI, Forskning 5:78.
Beeby A. W. *A note on various papers dealing with corrosion of reinforcement.* Cement and Concrete Association, Oct.
Beeby A. W. *Concrete in the oceans. Cracking and corrosion.* Construction Industry Research and Information Association *et al.*, Technical Report 1.
Beeby A. W. Cracking, what are crack width limits for? *Concrete*, No. 7, 31–33.
Bertrandy R. Corrosion à la mer de structures en béton armé et précontraint. *Technq. Bâtim. Trav. Publ.*, No. 30.
Blankenhorn P. R. *et al.* Chloride penetration of concrete impregnated with various linseed oil/mineral spirits combinations. *Cem. Concr. Res.*, **8**, 565–570.
Brown B. L. *et al. Corrosion testing of steels for reinforced concrete.* Building Research Establishment, Garston, CP 45/78.
Freisleben-Hansen P. *Hærdeteknologi. 1: Portlandcement. 2: Dekrementmetoden* (in Danish). Aalborg Portland og BKF-centralen.
Houghton and Paxton B. Cavitation resistance of some special concretes. *J. Am. Concr. Inst.*, **75**, No. 12.
Lenzner D. Stand der Forschung über die Wirkung von Alkalien in Zement und Beton. *Proc. 4th Int. Conf. Effects of Alkalis in Cement and Concrete.*
Rehm G. *et al.* Untersuchungen über den Korrosionsschutz von Spannstahlen unter Spritzbeton. *Dt. Aussch. Stahlbet.*, No. 298.
Rieche G. and Delille J. Erfahrungen bei der prüfung von temporaren Korrosionsschutzmitteln für Spannstahle. *Dt. Aussch. Stahlbet.*, No. 298.
Sprung S. Beton für Meerwasserentsalzungsanlagen. *Beton*, No. 7, 241–245.
Tuutti K. Cracks and corrosion. The corrosion of steel in concrete — the effects of cracks in the concrete cover. *Cem. Betong.*, 6:78.
Vassie P. R. W. *Evaluation of techniques for investigating the corrosion of steel in concrete.* Transport and Road Research Laboratory, Crowthorne, SR 397.
Vassie P. R. W. *The corrosion of steel in environments simulating the liquid phase of concrete.* Transport and Road Research Laboratory, Crowthorne, SR 396.

1979 American National Standards Institute and American Society for Testing and Materials. *Standard definitions of terms relating to concrete and concrete aggregates.* ANSI/ASTM, C125-79a.
American Society for Testing and Materials. *Standard definitions of terms relating to corrosion and corrosion testing.* ASTM, Philadelphia, G15-79a.
Arthur P. D. Fatigue of reinforced concrete in seawater. *Concrete*, **13**, No. 5, 26–30.
Arup H. Potentiotatic corrosion tests of steel in concrete for prolonged periods in fields and laboratory. *NACE corrosion meeting, Atlanta*, Korrosionscentralen Report 79/02c.
Arup H. and Grønvold F. Localisation of corroding reinforcement by electrochemical potential surveys. *Proceedings of symposium on quality control of concrete structures, Stockholm*, Report 79/04c, Korrosionscentralen. RILEM.
Beijer O. Weathering of concrete facades. In *Studies on concrete technology.* Swedish Cement and Concrete Institute, Stockholm.
Browne R. D. and Georghegan M. P. The corrosion of concrete marine structures. *Corrosion of steel in concrete construction.* Society of Chemical Industry, London.
CBI. *Corrosion of steel in concrete.* CBI Swedish Cement and Concrete Research Institute, Report of Activities 1978–79, 24–30.
Fagerlund G. Prediction of the service life of concrete exposed to frost action. In *Studies on concrete technology.* Swedish Cement and Concrete Institute, Stockholm.
Fagerlund G. Service life of structures. General report on sessions 2 and 3. *Proceedings of symposium on quality control of concrete structures, Stockholm.* RILEM.

FIP. *Report on prestressing steel*. 5: Stress corrosion cracking resistance. Tests on prestressing tendons. FIP/5/7, Draft.

FIP. *Report on prestressing steel*. 6: The influence of stray electrical currents on the durability of prestressed concrete structures. FIP/5/8, Draft.

Gjørv O. E. and Vennesland Ø. Diffusion of chloride ions from seawater into concrete. *Cem. Concr. Res.*, **9**, 229–238.

Huang C. L. D. Application of finite difference method to moisture transfer in porous media. *Proc. 1st Int. Conf., Swansea*, 539–547.

Huang C. L. D. Multi-phase moisture transfer in porous media subjected to temperature gradient. *Int. J. Heat and Mass Transfer*, **22**, 1295–1307.

Huang C. L. D. *et al.* Heat and moisture transfer in concrete slabs. *Int. J. Heat and Mass Transfer*, **22**, 257–266.

Knofel D. Einfluss von Frost und Taumittel auf Zementstein und Zuschlag. *Betonwerk Fertigteil-Technik*, **4**.

Lenzner D. and Ludwig U. Zum erkennen von Alkali-Kieselsaurereaktionen in Betonwerken. *Zem.-Kalk-Gips*, **32**, No. 8.

NCHRP. *Durability of concrete bridge decks*. NCHRP, Washington, Synthesis 57.

Nielsson L.-O. Moisture measurement in concrete by the hygrometric method. In *Studies on concrete technology*. Swedish Cement and Concrete Institute, Stockholm.

Nixon P. J. *et al.* The concentration of alkalis by moisture migration in concrete — a factor influencing alkali aggregate reaction. *Cem. Concr. Res.*, **9**, 417–423.

Organisation for Economic Co-operation and Development Road Research Group. *Evaluation of load carrying capacity of bridges*. OECD, Paris.

Ostlund L. Durability and safety. In *Studies on concrete technology*. Swedish Cement and Concrete Institute, Stockholm.

Pat M. G. M. Erosion of concrete. *Heron*, **24**, No. 3.

Rey E. Bewährung von Brücken in den Schweizer Alpen. *Schweizer Ing. Archit.*, 38/79.

RILEM. *Proceedings of symposium on quality control of concrete structures, Stockholm*, vols 1 and 2. RILEM.

Slater J. E. Criteria for adequate cathodic protection of steel in concrete. *Corrosion 79, Houston, Texas*, Paper 132.

Society of Chemical Industry. *Corrosion of steel reinforcement in concrete construction*. SCI, London.

Soretz St. Korrosion von Betonbauten — ein neues Schlagwort. *Zem.-u. Bet.*, No. 1.

Swedish Cement and Concrete Institute. *Studies on concrete technology*. SCCI, Stockholm.

Thistlewayte D. *Sulphide in Abwasseranlagen*. Melbourne and Metropolitan Board of Works, Betonverlag.

Tuutti K. Service life of concrete structures — corrosion test methods. In *Studies on concrete technology*. Swedish Cement and Concrete Institute, Stockholm.

Vassie P. R. *The influence of chlorides on the alkalinity*. Transport and Road Research Laboratory, Crowthorne, LR 915.

Weigler H. Entwicklung der Internationalen Betontechnologischen Bestimmungen. *Beton*, No. 6, 217–220.

Werse H.-P. Erfahrungen bei der Prüfung des Frost-Tausaltz-Wiederstandes von Beton. *Beton*, 12/79, 435–440.

1980 Alexyen S. and Rosental V. *Korrosion von Stahlbeton in aggressiver Industrieluft*. Berlin.

American Concrete Institute. *Performance of concrete in marine environment*. ACI, Detroit, SP-65.

Anderson G. H. Cathodic protection of a reinforced concrete bridge deck. *Concr. Int.*, **2**, No. 6, 32–36.

Andre D. Application du méthodes eléctrochimiques à l'étude du peintures anticorrosion. *Bull. Liais. Labs Ponts Chauss.*, Nov.–Dec., 110.

Bakker R. *Über die Ursache des erhöhten Widerstandes von Beton mit Hochofenzement gegen die Alkali-Kieselsaurereaktion und den Sulfatangriff*. Dissertation Aachen.

Beeby A. *Cover to reinforcement and corrosion protection*. Fédération Internationale de la Précontrainte, Notes 89, 4–8.

Calleja J. *Durability*. Instituto Eduardo Torroja de la Construccion y del Cemento (IETCC), Madrid.

Committee on Status of Cement and Concrete R&D in the United States. *The status*

of cement and concrete R&D in the United States. National Materials Advisory Commission on Sociotechnical Systems, Washington, D.C.
Crozier W. F. G. *FIP Commission on concrete sea structures*. FIP notes 87, 2–5.
CUR-VB. *Beton en afvalwater*. CUR-VB, Rapport 96.
CUR-VB. *Erosie van beton*. Concrete Society, Zoetermeer, CUR-VB Report 99.
Danish Ministry of Transport. *Concrete durability. An investigation of recently constructed motorway bridges in Denmark*. DMT.
Efes Y. Einfluss der Zemente mit unterschiedlichem Huttensandgehalt auf die Chloriddiffusion im Beton. *Betonwerk Fertigteil-Technik*, No. 4, 224–229; No. 5, 302–306; No. 6, 365–368.
Fagerlund G. *Influence of slag cement on the frost resistance of concrete — a theoretical analysis*. Swedish Cement and Concrete Research Institute, Stockholm.
Figg J. W. Rusting reinforcement — the No. 1 problem of concrete durability. *Concrete*, **14**, No. 5, 34–36.
Forschungsinstitut des Vereins der Osterreichischen Zementfabrikanten. *Grundsatzliche Untersuchungen an Stahl in alkalisch-Wassriger Losung bei Gegenwart von Chloriden als Modell für das Verhalten von Stahleinlagen im Beton unter Chlorideinwirkung*. FVOZ, Report P1/80.
Forschungsinstitut des Vereins der Österreichischen Zementfabrikanten. *Korrosionsschutz der Betonstahlbewehrung bei Tausalzeinwirkung Chloridkorrosion — Erste Auswertung des Schrifttum*. FVOZ, Report B442, 1.
Forschungsinstitut des Vereins der Österreichischen Zementfabrikanten. Korrosionsschutz der Betonstahlbewehrung bei Tausalzeinwirkung. *Forschungsvorhaben*, No. 977 im Auftag des Bundesministeriums für Bauten und Technik.
Fratteberg A. Frost action in mortar of blended cement with silica dust. *Proc. Am. Soc. Test. Mater.*, 536–548.
Frost-resistance concrete, international colloquium, Vienna. *Mitt. ForschInst. Ver. Öst. ZemFabr.*, No. 33.
Huberty J. M. *Durability d'aspects des betons apparents, le viellissement des façades*. C.S.T.C., F.I.C., FeBe.
Idorn G. M. Interface reactions between cement and aggregates in concrete and mortar-bond strength and durability. General Report, theme 7. *Proc. 7th Int. Congr. Chem. Cem., Paris*.
ISO/TC 71/SC 3/ad hoc group 1, no. 9. *Classification of environmental conditions affecting concrete*.
Kari A. *Investigation on the improvement of corrosion resistance and strength of ribbed reinforcing steel bars in concrete, especially when subject to fatigue loading*. Technical Research Centre of Finland Building Technology and Community Development, Publication 17.
Martin H. *et al.* Berechnungsverfahren fur Rissbreiten aus Lastbeanspruchung. *Forsch. StrBau StrVerkTech.*, No. 309.
Nurnberger U. Analyse und Auswertung von Schadensfallen an Spannstahlen. *Forsch. StrBau StrVerkTech.*, No. 308.
Peterson C.A. Survey of parking structure deterioration and distress. *Concr. Inst.*, **2**, 53–61.
Raharinaivo A. Utilisation à titre experimental d'armatures de precontrainte en acier galvanise pour la confortation par la precontrainte additionelle du Pont de Roquemaure. *Bull. Liais. Labs Ponts Chauss.*, Nov.-Dec., 110.
Rehm G. *et al.* Versuche zur Ermittlung der Korrosionsempfindlichkeit von Spannstahlen. *Forsch. StrBau StrVerkTech.*, No. 309.
Rostam S. and Pedersen E. S. Partially prestressed concrete bridges. Danish experience. *Proc. FIP symposia on partial prestressing and practical construction in prestressed and reinforced concrete, Bucharest*, Part 1, 361. Also: TU of Denmark, Department of Structural Engineering, Report Series R, No. 129.
Soretz S. Lernen aus Fehlern im Baugeschehen. *Zem. Bet.*, No. 4, 140–145.
Udvalget vedrørende betons holdbarhed. *Udredning om forskning og udvikling (F&U)*. (*An investigation into research and development related to concrete durability*.) Statens Teknisk-Videnskabelige Forskningsråd og Teknologirådet, Denmark.
Vassie P. R. *A survey of site tests for the assessment of corrosion in reinforced concrete*. Transport and Road Research Laboratory, Crowthorne, LR 953.
Volkwein A. Eindringen von Chloridionen in Beton. *Str. Autobahn*, No. 4, 168–173.
Weigler H. Beton — Anforderungen und Möglichkeiten. *Betonwerk Fertigteil-Technik*, No. 1080.
Weigler H. and Sieghart K. Structural lightweight aggregate concrete with reduced

density — lightweight aggregate foamed concrete. *Int. J. Lightweight Concr.*, No. 2, 101–104.

Wicke M. Erfahrungen aus der Inspektion von Massivbrucken. *Zem. Bet.*, No. 3.

1981 Bache H. H. Densified cement/ultra-fine particle-based materials. *2nd Int. Conf. Superplasticizers in Concr., Ottawa.*

Christensen P. *et al.* Filling up of pores and fractures in weathered concrete structures. *Proc. 3rd Int. Conf. Cem. Microscopy, Houston.*

Christensen P. *et al.* Microscopic studies of aggregates in Danish concrete. *Proc. 3rd Int. Conf. Cem. Microscopy, Houston.*

Gudmundsson H. *et al.* Quantitative microscopy as a tool for the quality control of concrete. *Proc. 3rd Int. Conf. Cem. Microscopy, Houston.*

Idorn G. M. Alkali-silica reactions — 1981 state of affairs. *Summary report 5th Int. Conf. Alkali-Aggregate Reactions, Cape Town*, Special publication.

Idorn G. M. International aspects of development of the uses of fly ash with cement. *Proc. Ann. Mtg*, Symposium N, Session 6. Material Research Society.

Idorn G. M. *Research and development for the use of fly ash in cement and concrete.* Electric Power Research Institute, Palo Alto, Technical Planning Study WS81-232.

Kosin F. Die Korrosion der Bewehrung durch elektrochemische Lokalelemente und Streustrome. *Str.-u. Tiefb.*, **35**, No. 6, 22–29.

Lawrence C. B. *Durability of concrete: molecular transport processes and test methods.* Cement and Concrete Association, Wexham Springs, Technical Report 544.

Lohrengel M. M. *et al.* Inhibitors of passive corrosion — application of a new test with thin oxide films. *Werkstoffe Korros.*, **32**, 13–18.

Organisation for Economic Co-operation and Development Road Research Group. *Bridge maintenance.* OECD, Paris.

Page C. L. *et al.* Diffusion of chlorideions in hardened cement pastes. *Cem. Concr. Res.*, **11**, 395–406.

Polster H. and Friedrich J. Neue Vorschriften zum Korrosionsschutz im Betonbau. *Bauplan. Bautech.*, No. 1.

Proc. 5th Int. Conf. Alkali-Aggregate Reactions, Cape Town.

Regan P. E. and Hamadi Y. D. *Behaviour of concrete caisson and tower members.* Concrete in the Oceans, Technical Report 4.

Regourd M. Resistance of concrete to chemical attack. *R.I.L.E.M. Matér. Construct.*, **14**, No. 80.

RILEM-Commission 42-CEA. Properties of set concrete at early ages. *R.I.L.E.M. Matér. Structs*, Nov.–Dec., No. 84.

Seki H. Deterioration of reinforced concrete wharf. *Concr. Int.*, **3**, No. 3, 57–65.

Sneck T. RILEM and durability. *R.I.L.E.M. Matér. Structs*, **14**, No. 83.

Sørensen B. *Korrosion af varmforsinket armeringsjern* (Corrosion of hot dip galvanized reinforcing steel). PhD thesis, Licentiatafhandling, Instituttet for Metallære, Danmarks Tekniske Højskole.

Teknologisk Institut, Byggeteknik. *Udsatte Betonkonstruktioners holdbarhed* (Durability of exposed concrete structures). Technological Institute, Denmark.

Vesikori E. *Corrosion of reinforcing steels at cracks in concrete.* Technical Research Centre of Finland, ESPOO.

Woodward R. J. *Case studies of the corrosion of reinforcement in concrete structures.* Transport and Road Research Laboratory, Crowthorne, LR 981.

Zaitsev Y. B. and Wittmann F. H. Simulation of crack propagation and failure of concrete. *R.I.L.E.M. Matér. Structs*, No. 83.

1982 Mather B. When will the concrete industry enter the 20th century? *Concr. Int.*, Jan., 25–28.

Rostam S. Structural concrete in Denmark. 9th FIP Congress, Stockholm. *Nord. Betong*, Nos 2–4, Special edition.

Thaulow N. *et al.* Estimation of the compressive strength of concrete samples by means of fluorescence microscopy. 9th FIP Congress, Stockholm. *Nord. Betong*, Nos 2–4, Special edition.